CHICKEN & EGG

An egg-centric guide to raising poultry

Andy Cawthray
&
James Hermes

Ivy Press

First published in the UK in 2015 by
Ivy Press
210 High Street
Lewes
East Sussex BN7 2NS
United Kingdom
www.ivypress.co.uk

British Library Cataloguing-in-Publication Data
A catalogue record for this book is available from the British Library

ISBN: 978-1-78240-200-8

This book was conceived, designed and produced by
Ivy Press
Creative Director: Peter Bridgewater
Publisher: Susan Kelly
Editorial Director: Tom Kitch
Art Director: James Lawrence
Editor: Jamie Pumfrey
Design: JC Lanaway
Paintings: Kate Osborne

Printed in China

Colour origination by Ivy Press Reprographics

10 9 8 7 6 5 4 3 2 1

Distributed worldwide (except North America) by Thames & Hudson Ltd, 181A High Holborn, London WC1V 7QX, United Kingdom

Contents

Preface

When considering chickens, the question always arises: "Which came first, the chicken or the egg?" This single conundrum has plagued the great thinkers of human civilization for millennia. Of course, it has no real answer because its purpose is simply to illustrate that some questions don't have answers, and yet the debate will continue long into the future. However, while the debate rages with little chance of any resolution, we have chickens and eggs to fill our lives with both pleasure and nutrition.

The chicken has become entwined in the lives of most of the world's population. In any restaurant or café, chicken and eggs are prominent fare on the menu. In knick-knack and general-household stores, chickens of all shapes and sizes are prominently on display, immortalized in stylized paintings and sculptures, and chicken feathers are found in many crafts, hats, pillows, and fishing flies. Chickens have also become part of our language, in sayings such as "Don't put all your eggs in one basket," "Don't count your chickens before they've hatched," "You've got egg on your face," "The hen-pecked husband flew the coop," and "Don't get your hackles up."

In recent years, the poultry and egg industries have become the primary producers of protein consumed by over half of the population of the world. The chickens used to produce these foodstuffs have developed into the most efficient traditional livestock species on the planet, returning the highest level of product—meat and eggs—for the least amount of feed.

In small farms and even in urban settings, chickens have gained in popularity. Small flocks of a handful of hens are supplying eggs to many families and their neighbors while providing life lessons, responsibility, and the hard realization of life and death to countless numbers of children.

This book will lead the reader through the origins and history of the birds we know as chickens. The complexity of the lowly egg is explained, including how it is formed, its quality criteria, and the conditions that result in baby chicks. Potential and new small producers will learn of the variety of egg-laying breeds and their temperaments, as well as the basic management conditions and practices required to house a small flock, including the proper care and feeding of these impressive animals.

We may not answer the questions as to why the chicken crossed the road or whether the chicken or the egg came first, but we will lead you on an exciting journey into the world of the chicken and the egg.

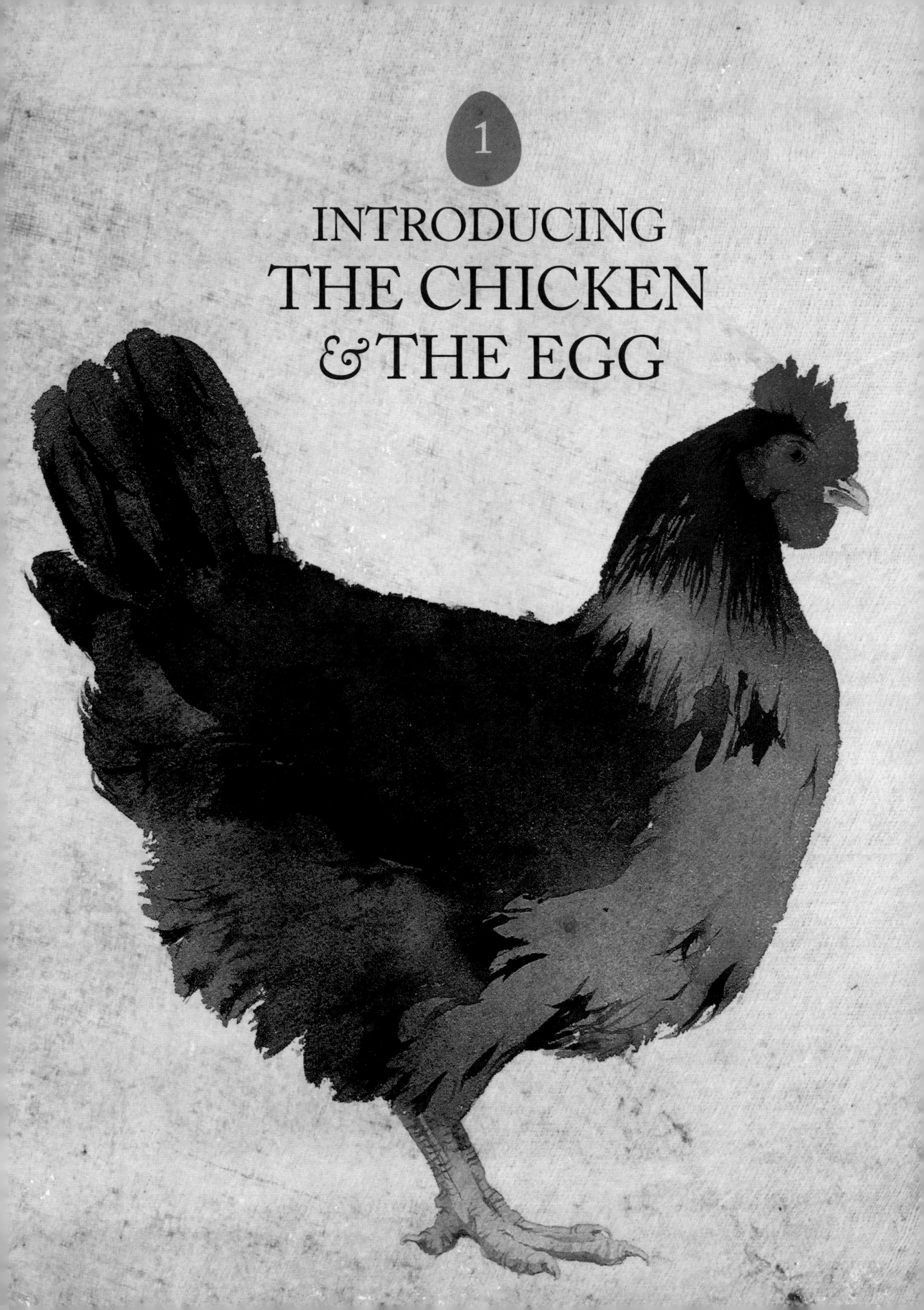

1

INTRODUCING THE CHICKEN & THE EGG

CHAPTER ONE

CHICKENS What Are They?

Chickens are a group of domestic birds that have been associated with man for millennia. Since their domestication for use as livestock animals, they have been a source of wonder and amazement, as well as becoming one of the primary protein sources in the human diet, in the form of chicken meat and eggs.

A Brief History

Chickens are essentially a domesticated pheasant native to Southeast Asia. They are classed with similar birds in the taxonomic order Galliformes (commonly called gamebirds or galliforms), which, in addition to the chicken and its wild progenitor, the jungle fowl, include turkeys, pheasants, quail, partridges, guinea fowl, and peafowl. Galliforms in general are adapted to life on the ground. While they can fly, none is a strong flyer, and most species use this ability for the primary purpose of avoiding predators; an explosion of wing-flapping to take off (the flush) startles the predator and lasts for only a few moments, with the bird landing a short distance away and the predator standing confused and without a meal.

In the beginnings of their association with man, early "pre-chickens" were relegated to a recreational role. It is generally accepted that jungle fowl were originally taken into captivity for fighting contests, with the winning roosters gaining the right to produce the next generation. As jungle fowl morphed into chickens during the ensuing millennia, fighting skills remained high on the priority list for selection. Egg and meat production were inconsequential during the domestication and early development of the chicken.

Right
Red jungle fowl, a member of the avian taxonomic order Galliformes, is the ancestor and progenitor of the modern domestic chicken.

Over the centuries, as chickens continued to be bred for their fighting ability, many of the traits that are important in breed identity today arose merely as the result of mutations brought about by selective breeding. Traits such as feather color and pattern, comb type, feathered shanks, polydactyly (extra toes), and skin, egg, and earlobe color were identified by these early chicken enthusiasts and selected for during breeding in addition to fighting skill. This selection of groups of birds with similar traits could be considered the beginning of breed development. It is suggested that the Dorking breed, which has the distinguishing trait of five toes, was described as early as the first century AD. However, it wasn't until the middle of the 19th century that chicken breeds, based on body shape and size, comb type and skin, and leg, egg, and earlobe color, among other traits, began to be categorized in an official way. At the same time, cockfighting began to lose favor with the public, with the sport outlawed first across most of Europe (it was banned in England and Wales in 1835), and finally in the US in 2006. While cockfighting remains a source of recreation in many parts of the world, chicken enthusiasts are increasingly turning to "fancy chickens," competing by exhibiting them in shows rather than allowing them to fight in pits.

Breed standards

In about 1863, the first Poultry Club of Great Britain was established, publishing the first standard of poultry breeds in 1865. The club was disbanded shortly after, only to be resurrected in 1877 to the current Poultry Club of Great Britain. In 1873, the American Poultry Association was founded; it published its first Standard of Perfection in 1874, which has been updated ever since, with a new edition every few years.

With the advent of showing, breed development intensified and breed standards were developed. Breeds were required to "breed true"—produce offspring similar to their parents—and had to conform to specific standards (see box) for weight, shape, color, and many other traits.

From their domestication until the early years of the 20th century, chickens were not an important part of the human diet. Birds were bred for fighting or their appearance, and only culls or hens past their prime were consumed.

Animal production versus feed required

Eggs were readily consumed but were not a large part of the diet either. However, a small segment of poultry breeders began to disregard breed characteristics in favor of economic traits such as egg number, size, and quality, or meat traits such as growth rate, body size, and feed conversion.

Commercial egg production began as a few extra birds laying eggs for sale or barter. Flocks rarely exceeded a couple of hundred birds, and hens laid only a few dozen to a hundred or so eggs annually. In 1913, after several years of selective breeding for egg number, James Dryden, professor of poultry husbandry at Oregon State University, produced Lady McDuff, the first hen that was recorded to lay more than 300 eggs in a single year. A few years later, Dryden bred Oregona, a hen that was documented as laying more than 1,000 eggs in five years. Both hens were white-egg-laying single-comb white Leghorns and marked the beginning of the egg-laying strains of chicken in the modern era.

Breeding chickens specifically for their meat didn't take off until as recently as the 1920s, when Cecile Steele first raised chickens for the purpose in Delaware. However, the quality remained poor until Charles Van Tress of California crossed Dark Cornish and New Hampshire breeds, selected for their growth and meat characteristics. The result, the Cornish Cross, won the A&P supermarket chain's Chicken of Tomorrow contest in 1948 and gave rise to the modern broiler chicken.

From these beginnings, through many generations of selection and further crossing with the white Plymouth Rock to bring in white feathering, the Cornish Cross remains a non-standard breed. Nonetheless, it has ideal meat characteristics of growth and feed conversion, and to the present day remains the most efficient traditional livestock species, growing to nearly 7 lb. (3.2 kg) in as many weeks while consuming about 12 lb. (5.4 kg) of feed.

In recent years, chicken has become the primary meat of choice for much of the western world, and with the addition of eggs in our diet, the bird itself has arguably become the most important livestock animal throughout the world.

Feed conversion

One way to determine the efficiency of production of livestock is by comparing the amount of feed consumed with the product produced, either meat or eggs.

Origins of the Chicken

The chicken, which is arguably the most numerous bird on the planet, began its existence as an obscure pheasant in the jungles of Southeast Asia. It was most probably Charles Darwin who first proposed that the bird was the domestic ancestor of the red jungle fowl (*Gallus gallus*).

This species is among a group of 4 extant and as many as 13 extinct pheasant-type species that are found from western India through the southern reaches of Asia and into the island chains of Indonesia and the Philippines. The birds generally dwell on the jungle floor and into the margins of grasslands at the edge of forested areas in these regions. The ranges of these closely related species rarely intersect, so there is little possibility of natural hybridization between them.

The red jungle fowl, which has the broadest range—from India in the west through Southeast Asia and into the many of the

Comb characteristics

Both the Sri Lankan and green jungle fowl vary in appearance from the more common red and gray jungle fowl, especially in terms of their comb. The combs of these more isolated types are variable in color, with not only the typical red but also patterns of blue and yellow. In addition, the Sri Lankan jungle fowl's comb has even smaller and less distinct serrations when compared with that of the gray jungle fowl, and the serrations are completely lacking in the comb of the green jungle fowl.

Right
Examples of the four extant species of jungle fowl that are found in the wild in Southeast Asia, from western India to the east coast of China and into the islands of Indonesia.

Left
Combs of jungle fowl and chickens have distinctive morphology and are important for breeding displays and probably thermoregulation.

Indonesian and Philippine islands—has at least five subspecies groups that are somewhat isolated from one another. In appearance and behavior, it most resembles the domestic chicken; in fact, many novice and even some experienced poultry enthusiasts could mistake it for a chicken. Truly wild examples of red jungle fowl are probably difficult to obtain today because the birds have been breeding with domestic chickens for centuries as their habitat has been infiltrated by man.

The other jungle fowl species have relatively small ranges and are mostly isolated. The gray jungle fowl (*Gallus sonneratii*), sometimes called Sonnerat's jungle fowl, is found in the westernmost part of the overall range of jungle fowl, generally in peninsular India. It differs most strikingly from its red jungle fowl cousin in its comb—the flesh appendage that resides on the top of the head, used for courtship display and probably for cooling. Most prominent in the males, the small points and minimal serrations of the gray jungle fowl's comb are clearly in contrast with the more chicken-like "single comb" of the red jungle fowl. (The single comb is the typical comb of most breeds of chicken; it mostly stands erect with "sharp" points forward and on top, and a broader "blade" projecting rearward.)

The remaining two jungle fowl species are isolated to some of the islands that lie within the overall range of the *Gallus* genus. The Sri Lankan jungle fowl (*Gallus lafayetii*; previously called the Ceylon jungle fowl) is found only on the island of Sri Lanka off the eastern tip of southern India, while the green or Javan jungle fowl (*Gallus varius*) is found on a few islands in western Indonesia, including Java and Bali.

Domestication of the Chicken

While the exact date and place of chicken domestication is impossible to determine, it is clear when comparing the genetic evidence from domestic chickens and jungle fowl that the vast majority of the chicken genome is quite similar to that of the red jungle fowl and dissimilar to the other extant *Gallus* species. Genetic studies also reveal that there were probably two major domestication events: one in the Indus Valley of western India, and the other in Southeast Asia, in the region of Vietnam.

In 2008, following the completion of the sequencing of the complete chicken genome in 2004, it became clear that at least some of the DNA from the gray jungle fowl had made its way into the chicken. The gene that imparts yellow color to a chicken's skin, which is a common trait in many chicken breeds, is found only in the DNA of the gray jungle fowl. It is unknown whether the addition of this single known gene (more may be discovered in the future) occurred prior to domestication as a natural hybridization or as a result of "domestic" red jungle fowl breeding with gray jungle fowl sometime after the original domestication event.

The domestication of the chicken was apparently contemporary with the domestication of other major animal species. The dog was arguably first in the domestication parade, followed by pigs, cattle, sheep, goats, horses, and, much later, turkeys, with chickens coming on the scene around 8,000 years ago, between the times goats and horses were domesticated.

While food and fiber were probably of primary concern for the domestication of most traditional animal species, it seems that the chicken was originally domesticated for the propensity of the roosters to fight. It is surmised that early man watched red jungle fowl roosters fighting off other roaming roosters in short but dramatic wing-flapping battles, leading to the banishment of one and the reproductive success of the other.

Fighting is a natural behavior of the polygamous red jungle fowl. The birds are harem breeders, meaning that each rooster will select a string of between 8 and 12 hens to constitute his harem. He protects his harem from predators and from marauding stag roosters looking for a harem of their own. This natural process assures that only the strongest males produce the next generation.

Once these birds were captured and forced to live in captivity, their ultimate fate was assured—domestication. Over the succeeding centuries, populations of jungle fowl/chickens made their way north to China, arriving there

as early as 5,000 BC, and west into the Middle East, Europe, and Africa, which they reached by the first century AD. The birds also moved east, island-hopping throughout the Pacific Ocean with the Polynesians, until they made their first landfall in the Western Hemisphere in the 12th century on the coast of South America. At some point during their travels, jungle fowl became chickens.

Above

Chicken fighting has had a significant role in chicken development for millennia and remains important in some cultures in modern times.

CHAPTER TWO

The Science of Egg Formation

Egg-laying (oviparity) is a reproductive strategy in many animals. Of the mammals, only the platypus and four known species of spiny anteater (echidna) are classed as monotremes (egg-laying mammals), but among the reptiles, amphibians, and fish, many species are egg-layers. The eggs of all of these non-avian egg-laying species have a soft shell, which is either leathery or coated in jelly. Only bird eggs have a hard shell. The question many people ask is why do chickens lay eggs, or, more appropriately, what is the stimulus that causes them to lay eggs?

Reproductive Maturity

Eggs are the natural response of birds—in this case chickens—to reproduce. In general, pullets (young hens less than one year old) begin laying eggs at about 20 weeks of age and will continue to lay well for a year or so. As they age, their egg production begins to subside, increasing in the spring and summer and reducing in the fall and winter. But why do hens begin laying at 20 weeks?

As with all animals, age is a critical factor for reproduction. Maturity of the reproductive system, as well as of other organ systems and even a minimum body weight, are essential for successful reproduction. In mammals, puberty is attained at a certain age, which differs depending on the species. When puberty is reached, the brain signals the hypothalamus (the "master gland") and pituitary to release certain hormones directly into the bloodstream. When these hormones reach the gonads (the ovary in females and testis in males), they respond by maturing into their functional reproductive state.

Days of the estrous cycle

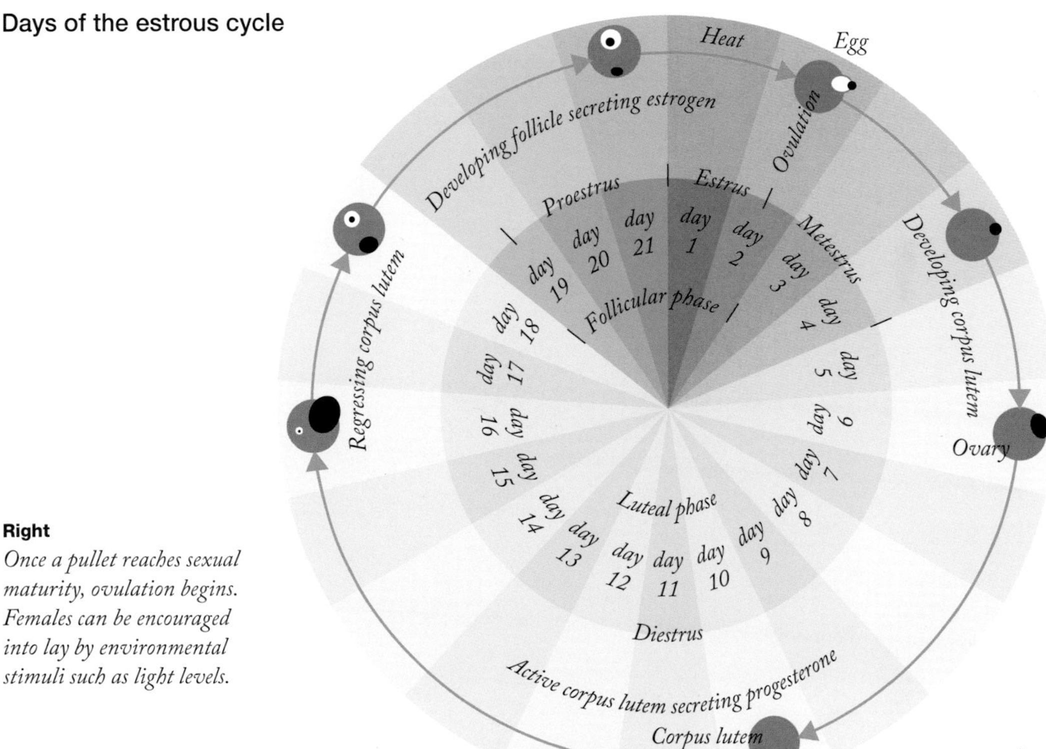

Right
Once a pullet reaches sexual maturity, ovulation begins. Females can be encouraged into lay by environmental stimuli such as light levels.

In female mammals, the ova that migrated to the ovary during embryonic development begin to complete their maturation, a process that occurs either one or a few at a time, depending on the species. In addition, the ovary releases hormones of its own, which target other sexual organs and trigger morphological changes to assure a successful reproductive event. After a few weeks, an ovum is released to meet with sperm from the male, which have undergone a similar maturation event. If fertilization is unsuccessful, the hormone levels subside for a period of about three to four weeks, when the process repeats. This is a simplistic review of the estrous or heat cycle in mammals.

THE STIMULUS: LIGHT

In chickens, and most birds in general, the process is similar but with a few dramatic differences. The first and most obvious difference is that chickens do not undergo puberty. Instead, it is a specific stimulus—the photoperiod, or light/dark cycle—that begins the process. The hen's brain remains the controlling organ, but in this case the pineal gland is key. The pineal is directly light sensitive (it is sometimes called "the third eye"), responding to changes in day length and controlling the circadian rhythm by releasing the hormones melatonin and serotonin. The hen will begin laying eggs when the day length reaches the threshold level of 12 hours.

When this level is reached and the hen is receiving at least 12 hours of light per day, its pineal gland is stimulated to release hormones that begin the cascade of events through the hypothalamus and pituitary and ending in egg production. For young pullets, the first egg will be laid after about two to three weeks, while older hens that have laid in the past will respond more quickly.

A second obvious difference with mammals is that rather than having an estrous cycle lasting three to four weeks, the hen will continue to lay eggs essentially daily for months while the stimulus of long day length remains. The pause in chickens occurs during fall each year, when the days become shorter. When day length drops below 12 hours, pineal hormones change and egg production slows and often ceases. During this lull in egg production, the hen will also molt, a process of losing and regrowing feathers. This natural process is much more severe in wild birds, which completely cease egg production in late summer and replace all their feathers by mid-fall. Some waterfowl find themselves flightless during the molt. Following centuries of breeding, the domesticated chicken often continues to lay during the molt and replaces only a portion of its feathers each year.

By taking advantage of the hen's biological response to light, farmers are able to increase substantially the number of eggs laid. Placing 18–20-week-old pullets under 14–16 hours of light per day will bring them into production in about two weeks, no matter what the season of the year. In addition, keeping the lights on a long-day schedule will keep them laying through the fall and winter months. Eventually the birds will need to molt, which can easily be accomplished during their second winter. This is done simply by providing them natural short-day photoperiods, starting in about mid-November and continuing through the end of December. This period of short days (less than 12 hours) will give the hens plenty of time to molt and rejuvenate their egg production. Restoring long days of 16 hours of light (and even a little more, as they are older now) from January onward will restore their egg production.

Left
Commercial egg production has begun to move from cage production to the use of aviary systems in which hens are confined for sleep and egg laying and allowed access to the floor during most of the daylight hours.

What about the rooster?

The question often arises: What about the rooster? What is his role in egg production? Unlike mammalian males that breed with the females only when they are in heat, the rooster, like the hen, responds to photoperiod and will breed with hens daily as long as the stimulus of long day length is present. So, while the rooster is obviously important for reproduction, he has no role in egg production.

Egg Formation: The Process

Biologically, eggs are the female's contribution to the reproduction process. Chickens, like all birds, produce a hard-shelled egg in which development of their offspring occurs without any outside contribution throughout the 21-day process.

The egg begins as a germ cell that has resided in the hen's ovary since the time the hen was an embryo developing in its own egg. Once a pullet reaches sexual maturity, ovulation begins.

As previously explained, birds do not experience puberty but are instead stimulated into sexual maturity by an environmental stimulus—photoperiod in temperate regions. Long days,

Reproductive system of the hen

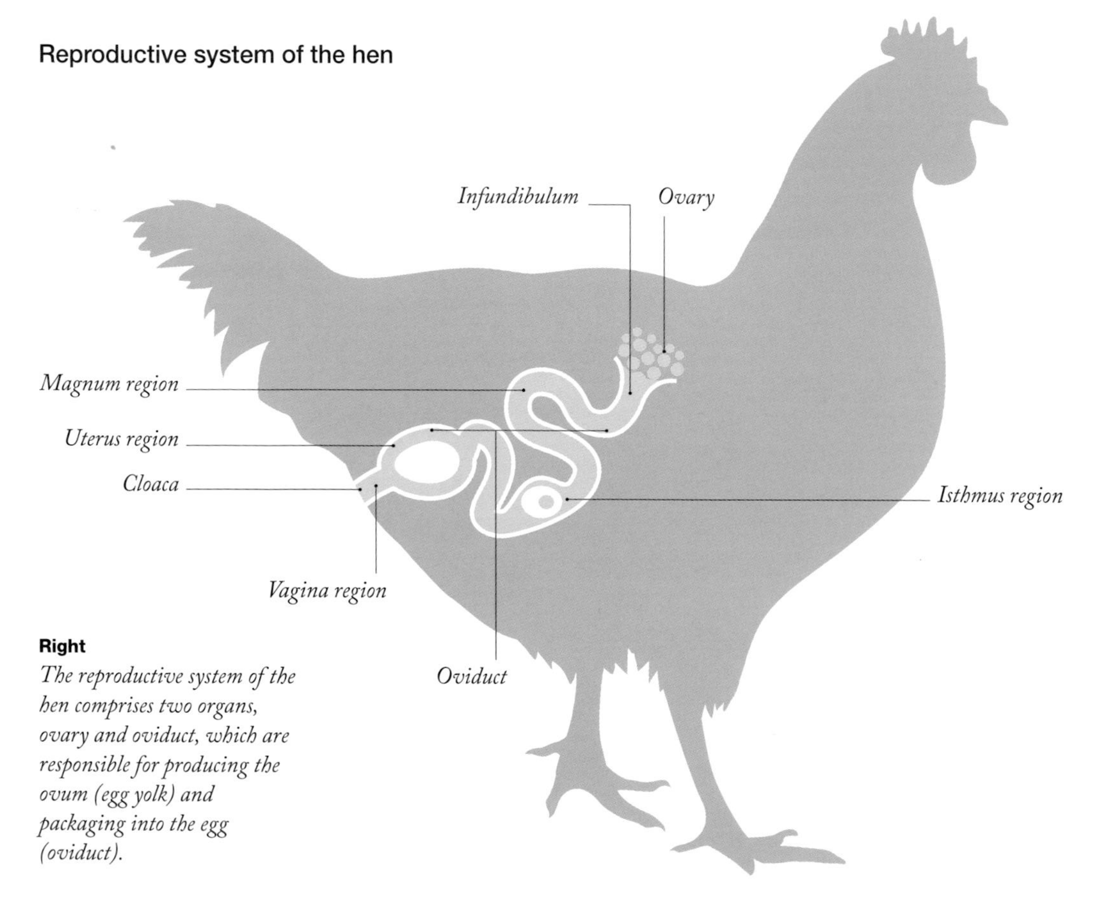

Right
The reproductive system of the hen comprises two organs, ovary and oviduct, which are responsible for producing the ovum (egg yolk) and packaging into the egg (oviduct).

with more than 12 hours of light, stimulate the brain and pituitary gland to release hormones that in kind stimulate the ovary to begin the process of ovum (yolk) formation. From the ovary, the formation of the egg passes through several stages, described as follows.

OVARY

Chickens have only one active ovary, always on the left. This contains as many as 3,000–10,000 germ cells, only a few of which develop at any one time. The germ cells that will eventually become yolks develop in order, so that only one is mature enough for ovulation (release). Occasionally, more than one yolk ovulates at a time, resulting in a double-yolk or multiple-yolk egg. Once the yolk is released from the ovary, the process of egg formation begins in the hen's oviduct.

OVIDUCT

As with the hen's ovary, only the left oviduct is active. Occasionally a "right oviduct" develops, but it is nearly always vestigial and non-functioning. The oviduct's function is essentially to package the yolk into what we know of as a chicken egg. The oviduct has five distinct sections, each with a function in the formation of the completed egg: the infundibulum, magnum, isthmus, shell gland, and vagina.

Infundibulum

The released yolk enters the oviduct through the infundibulum, the funnel-shaped end that actively engulfs the yolk. If the hen has mated in recent days, sperm cells that have been stored in "sperm nests" found in the infundibulum are released so that fertilization can occur. The yolk moves through the infundibulum in about 20 minutes.

Magnum

From the infundibulum, the yolk enters the magnum, passing through it in about three hours. It is in the magnum that albumen (egg white) is secreted around the yolk. Albumen is a combination of about 13 known proteins and is deposited around the yolk in distinct layers (see pages 25–6).

Isthmus

The yolk, which is now surrounded with albumen, then reaches the isthmus, a distinctly thinner and shorter part of the oviduct compared to the magnum. It is here that with the shell membranes, inner and outer, are deposited onto the albumen. During the nearly 90 minutes that the egg spends in the isthmus, water begins to diffuse into the albumen, plumping the overall package and beginning the formation of the ultimate egg shape. The final process in the isthmus occurs as the egg begins to move to the shell gland, and involves the placement of small "seeds" of calcium carbonate (calcite) onto the outer surface of the outer shell membrane. These calcite "seeds" provide the starting point for shell deposition in the shell gland.

Shell Gland

Nearly five hours after ovulation, the nearly formed egg passes from the isthmus into the shell gland. At this stage its overall shape and size have already been partially determined. The egg is still somewhat flaccid, soft, and pliable, yet completely covered by membranes, so if it were to be laid at this time, it would be deemed a "soft shell."

As the flaccid egg enters the shell gland, the plumping that started in the isthmus continues until the membranes are distended and the egg

Uterus versus shell gland

Many books continue to label the shell gland the "uterus." The use of this name has, unfortunately, continued for decades, with seemingly only avian purists having difficulty with its connection to an avian system. The shell gland section of the avian oviduct is not a uterus. A uterus is a mostly mammalian organ, where mammalian embryos implant, receive nourishment by way of a placenta, and develop from a few cells to a fully formed baby mammal, until "forced" out during parturition (birth). In a bird, the section erroneously called the uterus but more properly termed the shell gland has the biological function of secreting mainly calcium carbonate to create the hard eggshell, a distinctly avian characteristic.

no longer appears flaccid. During the next 20 hours or so, the rather thick lining of the shell gland secretes about 0.3 oz. (10 g) of calcium carbonate onto the membrane surface, "growing" calcite crystals to a thickness of about 0.007–0.011 in. (0.2–0.3 mm). The crystals are deposited in columns with occasional spaces left vacant, resulting in a porous shell. It is estimated that the pores make up about 0.003 in.^{2} (2 mm^2) of the total surface area of the shell of an average chicken egg. These pores allow exchange of oxygen, carbon dioxide, and water vapor.

Most breeds of chicken lay eggs with brown shells, a few lay white-shelled eggs, and only a couple lay eggs with blue or green shells. The brown color is deposited on the shell surface as one of the last events in the shell gland. The production of white-shelled eggs is due to a mutation in the biochemical pathway that creates the brown color, disrupting this so that the color is not deposited. The blue or green color of eggs from the Araucana and related breeds comes from the deposition of a different pigment that is "mixed" with the calcium carbonate shell so that the color is throughout the shell material. The difference is evident by examining the inside of the eggs after breaking the shell: Brown eggs are white on the inside surface, while blue and green eggs are blue or green all the way through. The entire process of shell formation takes about 18–22 hours.

Vagina

Once the shell is complete, egg-laying (oviposition) begins. Just prior to laying, the egg, which has been moving through the oviduct pointed-end first, turns so that the egg will be laid blunt end first, a movement that probably relates to ease of laying.

During the process of egg-laying, the egg moves to the vagina at the end of the oviduct, which is connected to the urodeum section of the cloaca, the end of the digestive tract. In a move that is thought to be a way of keeping the eggshell surface clean, the hen's oviduct everts from the cloaca, essentially turning "inside out." As the egg moves through the vagina, a coating of mucus is laid down that serves to protect the eggshell pores from bacterial contamination and to reduce water loss from the egg after laying, which preserves egg quality.

The sound of success

In most hens, hormones are mobilized within 20 minutes of laying that stimulate the hen to cackle, telling the world she has just produced an egg, as well as ovulating the yolk for tomorrow's egg when she does it all over again.

The Parts of the Egg

Eggs are a common sight in most kitchens, and while they may seem common, nature has taken great pains to create the avian egg into a powerhouse of nutrition, packaging it in a virtually sterile, single-serving container that makes most food scientists envious.

ALBUMEN

Albumen, or egg white, nearly exclusively comprises protein and water. Water makes up about 85 percent of the total albumen volume, with the remaining portion being mostly protein—about half of the total protein in the egg. The final 2–3 percent of the albumen comprises some minerals and a small amount of carbohydrates in the form of glucose and glycoprotein.

No fewer than 13 different proteins are found in the albumen, the main one (making up more than half of the total albumen protein) being ovalbumin, with the other major types including conalbumin (ovotransferrin), ovomucoid, ovomucin, and lysozyme. These proteins have various roles in nutrition as well as maintaining the thickness of the albumen, binding metals and other compounds, and keeping potential bacterial pathogens in check.

The combination of these 13 proteins and their constituent amino acids has been found to be ideal for human consumption. When measuring the human requirement for each of these amino acids, egg albumen and yolk are considered the best source for the human diet.

Albumen is deposited around the yolk in four distinct layers: the inner and outer thick, and the inner and outer thin. Directly surrounding

Avidin and biotin deficiency

One protein, avidin, which is found in small amounts in albumen, has the ability to "bind" the essential vitamin biotin. While unlikely, eating many raw eggs could result in a biotin deficiency. Cooking denatures the avidin slightly, so its ability to bind this important vitamin is eliminated.

Parts of the egg

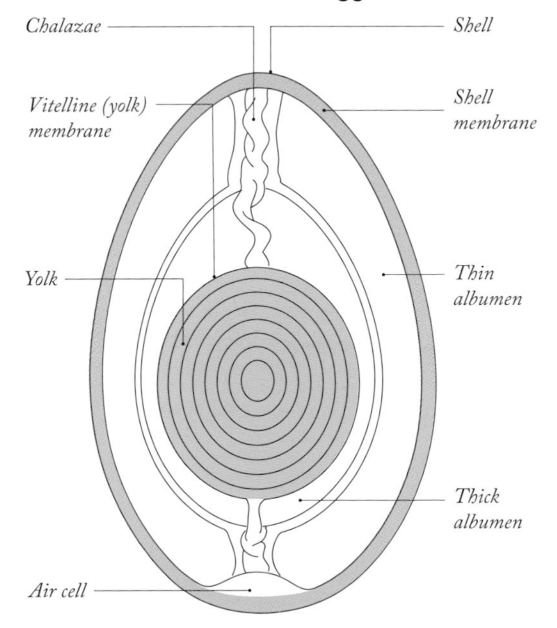

the yolk and the first layer to be deposited is the inner thick, also called the chalaziferous layer. This layer constitutes about 3 percent of the total and is observed as the nearly opaque, ropey structures (the chalazae, singular chalaza) on opposite sides of the yolk when the egg is broken open. When the egg is intact, this thick layer provides stability to the yolk, restricting its movement and preventing it from floating around the albumen. As the egg ages and the albumen becomes thinner, the chalazae lose their ability to hold the yolk in the center of the egg. Prominent chalazae are indicative of a fresh egg.

The inner thin layer, which could be described as watery, comprises nearly 25 percent of the total albumen but is not detectable and has no obvious function. The outer thick layer comprises over half of the total albumen. It is this layer that is observed during egg grading for firmness. In fresh eggs, the outer thick layer provides support to the yolk to make it stand tall. As the firmness is lost over time, it is this layer that is obviously responsible for downgrading. The final layer, the outer thin, is produced during the plumping stage of egg formation: As water from the oviduct diffuses into the as-yet-unformed egg during and following membrane formation, it mixes with the protein, creating this watery "slurry."

Typically, albumen is clear or slightly cloudy. Fresh eggs contain higher levels of dissolved carbon dioxide, resulting in a cloudy albumen. As the egg ages, the carbon dioxide is released through the porous shell, and the albumen becomes clear. Also noticeable is a slight yellow hue to the albumen, indicative of the vitamin B2 (riboflavin), which is at fairly high levels in eggs.

YOLK

Egg yolk, unlike the albumen, is fairly low in water content—it contains about 50 percent water when the egg is laid. However, because of the high water content of the albumen and the characteristic that water at high concentration will diffuse toward a lower concentration, the yolks slowly increase in water content during storage.

About half of the material in yolk is solids, primarily made up of proteins and lipids (fats). About half of the total protein of an egg is found in the yolk. The yolk proteins differ from those in the albumen and are found in combination with other compounds, yielding glycoproteins and high-density lipoproteins. Some of the proteins make up a protein solution that acts as a suspension material for other yolk particles of various structures and sizes. The larger of these particles are generally suspended in the protein solution, while the smaller particles are more or less free-floating. When observing yolk under magnification, tightly packed granules give it a yellow appearance, and when they are more loosely distributed, the yolk looks white.

Interestingly, during yolk formation in the ovary, over the course of about 10 days or more, the yolk material is deposited in layers—white and yellow yolk—that remain after laying. In fact, when egg yolks are stained under certain conditions in the laboratory, distinctive layers can be identified, and if fat-soluble dyes of various colors are fed to the chicken sequentially, a different color each day, each layer will reflect the color of the day. In addition, extending from the yolk's core (the point just under the undeveloped embryo) is a loose structure of white yolk called the latebra, which because of its lower density causes the

yolk to turn so that the embryo is always on the top.

The lipid fraction of yolk makes up about 35 percent of the total volume. These lipids comprise about 66 percent triglycerides, 29 percent phospholipids, and 5 percent cholesterol. The fatty acids that make up these lipids vary according to the feed intake of the hen. It is this fraction of the egg that can be greatly influenced by the hen's diet. For example, feeding high levels of omega-3 fats, as found in flax, can increased the content of omega-3 fatty acids in the egg. Overall, about 66 percent of the lipids are unsaturated fats, with the remaining 34 percent saturated fats. In recent years, research has shown that the fat and cholesterol content of eggs in the human diet has little effect on blood cholesterol levels.

Unlike albumen, whose pH increases after the carbon dioxide it contains diminishes after laying, there is little change in the pH levels of yolk. Yolk pH rises from about 6.0 to about 6.3 during storage (in comparison, albumen pH increases from around 7.5 to 9.0 or higher), and there is no apparent effect on the yolk.

Typically, egg yolks range in color from pale yellow to bright orange. These colors are the result of the hen's consumption of a group of carotenoid pigments called xanthophylls. The variation of the color is heavily based on the hen's diet (see box).

MEMBRANES & SHELL

The shell membranes are a minor part of eggs and not particularly remarkable, but they do play an important role in the egg as the base on which the shell can form. The inner and outer shell membranes, composed of protein fibers, are distinct microscopically in their appearance and texture. The inner membrane, which is in direct contact with the albumen, is composed of thin, fine fibers, while the fibers making up the outer membrane are comparatively thick and coarse.

The shell is composed of nearly 95 percent calcium carbonate, along with small amounts of calcium phosphate, magnesium carbonate, and protein. It is thought that the minor compounds impart hardness to the shell material and the protein provides a matrix to hold the calcium-carbonate crystals together as they "grow" into a ridged structure.

The most obvious characteristic of the shells of chicken eggs is that they can be found in various shades of brown, blue, green, and white. The natural color of chicken eggs, and the

Diet & yolk color

When hens consume feeds such as corn and alfalfa, or grasses and insects that are also high in xanthophylls, the eggs they produce have dark orange yolks. When highly pigmented feed sources are not available, dark orange yolks will soon fade to pale yellow. In those cases, processed feed additives like extract of marigold petals can be fed to darken the yolk once again.

Yolk color from a diet high in xanthophylls

Yolk color from a diet low in xanthophylls

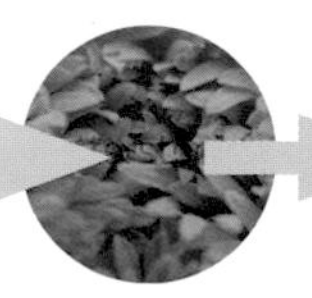

Processed feed additives added to diet

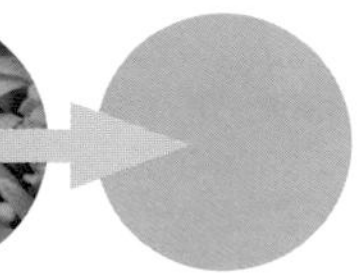

Yolk darkens again to dark orange

color produced by most breeds, is brown, a remnant of their jungle-fowl heritage.

The brown color of eggshells is caused by a group of pigments called ooporphyrins, which are synthesized using hemin and biliverdin, remnants of old blood cells. This color is deposited just prior to egg-laying so is on the outer surface only—the inside surface of brown eggs is white. The brown color can be removed with sandpaper to reveal the white shell underneath, and when the egg is still wet from laying, the color can sometimes literally be rubbed off.

White eggs are the result of a mutation in the chemical pathway synthesizing the ooporphyrin pigments. Most white-egg breeds, including the Leghorn, can be traced back to the Mediterranean region.

Green and blue eggshell colors are the result of another genetic mutation. It is thought that many centuries ago, a mutation took place in chickens that were centered on the west coast of South America, resulting in the modern Araucana breed, which lays green or blue eggs. When the gene responsible for the eggshell color is bred into chickens that normally lay brown eggs by crossing individuals of the two breeds, the resulting offspring will lay green eggs. However, if it is bred into chickens that lay white eggs, the eggshell color of the resulting offspring is pastel blue.

It must be noted here that eggshell color has nothing to do with the nutritional value, taste, or quality of eggs. It is simply that different breeds of hens lay eggs with different-colored shells based on their genetics.

Color research

University of Nottingham researchers, publishing in 2013, suggested that the blue/green eggshell gene sequence is not found in jungle fowl. So it is actually a new gene, possibly introduced into the chicken genome by a retrovirus. More research is needed to determine the validity of this work.

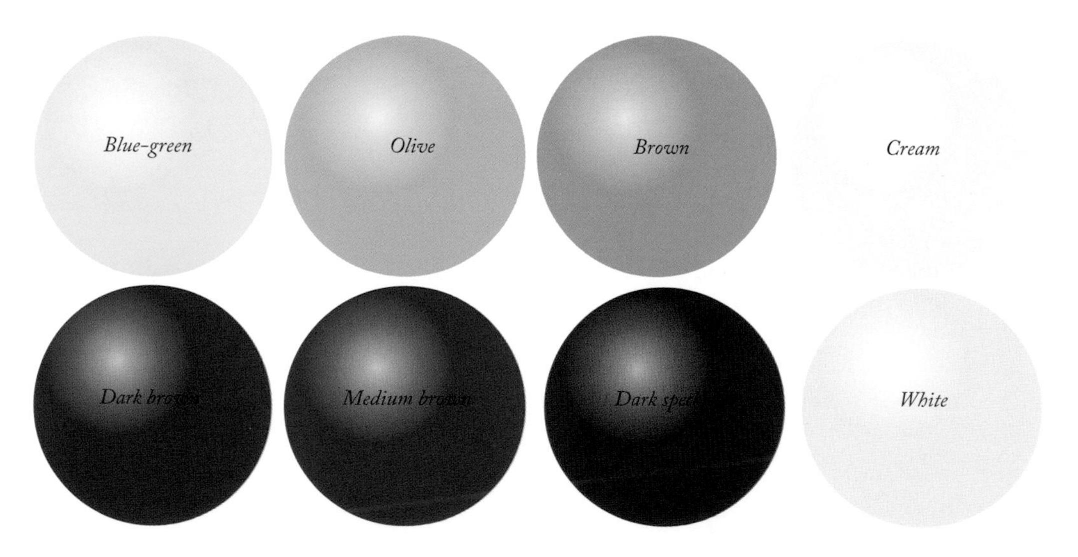

Chicken Reproduction

While the presence or absence of a rooster in the chicken flock has nothing to do with whether the hens lay eggs or not, he is essential for fertile eggs. As in all higher animals, reproduction in chickens is accomplished by the contribution of genetic material in the form of sperm and egg from both males and females, respectively. Sexual reproduction, compared with the asexual reproduction of many lower forms of life, provides for enhanced genetic variation within and between generations. Each offspring has a unique genetic "fingerprint," making it just a little bit different from all others. The result of this genetic variation is obvious when considering the multitude of breeds and varieties of the chicken.

Egg-laying is the process that birds, including chickens, use for reproduction. Eggs that hatch into chicks are formed in the same way as the eggs we eat, with just one minor difference: the introduction of sperm cells into the process that leads to a chick. During mating, the rooster deposits sperm cells into the hen's oviduct in a process called a "cloacal kiss." Roosters have a rudimentary copulatory organ, the papilla, which is used to transfer the sperm cells into the hen's system. Upon their arrival in the oviduct, the sperm cells reside for a brief time in "sperm nests," folds of tissue at the junction of the shell gland and vagina. Once the current day's egg is laid, these "stored sperm" move, with help from the oviduct, to the junction between the infundibulum and magnum, where more sperm nests are present. Once here, sperm are available for up to two weeks to join with any and all ovum (yolks) that pass, often resulting in fertilization—the union of sperm and egg.

Development of the chick begins immediately, with the first cell divisions (cleavages) occurring about four hours after fertilization and continuing until the egg is laid. At oviposition, the embryo has between 20,000 and 60,000 cells, giving the characteristic 1/8–3/16-in. (3–4-mm) doughnut or bull's-eye appearance to the germ, the developing embryo on the surface of the yolk. Once laid, the egg begins to cool. When the temperature reaches about 80°F (27°C), known as "physiological zero," cell division in the young embryo ceases and the embryo enters a period of quiescence.

Egg Incubation

Under natural conditions, chickens are indeterminate layers, producing 8–12 eggs in a clutch, at which time they become broody and "set"—nature's hormonal message for the hen to incubate her eggs. Hens begin to incubate the clutch after the last egg is laid, thereby hatching all of their young within a 24-hour period or so.

When artificially incubating eggs, we can take advantage of this natural pause in development to schedule setting and subsequent hatching. Following many years of research, the proper storage conditions for incubating eggs to maximize their hatching success have been determined (see page 31).

SETTABLE EGGS

When collecting eggs for hatching, note that some eggs should not be set. Eggs that are misshapen; have obviously poor-quality shells, thin spots, or a rough sandpaper-type surface; or are dramatically larger or smaller than the average in the group, will have poor hatching success and should be consider unsettable.

Dirty eggs should be cleaned or discarded. While washing by soaking or rubbing with a damp cloth is unnecessary, a quick rub with some fine sandpaper to remove any dried dirt or feces will reduce the possibility of contamination, which can result in infected eggs and problems later during incubation. Washing removes the cuticle, which is the hen's protective barrier to contamination. Eggs that are excessively dirty should not be set.

INITIAL STORAGE

It must be stressed here that eggs should not be set immediately upon laying. Setting eggs that are still warm from the hen tends to result in reduced hatching success. The biological reason for this phenomenon is not fully understood but the results are well documented. Therefore, setting should take place no earlier than after about two days of storage. The maximum storage period that still results in a high hatchability rate depends on the conditions in which the eggs are kept, including temperature, humidity, position, and whether the eggs are turned.

Temperature during Storage

As with many factors affecting biological development, time and temperature are closely related. Therefore, when it is considered how long hatching eggs can be stored, the next question is immediate: What is the temperature of that storage?

Research tells us that hatchability of stored chicken eggs is best when the eggs are kept at cool, not cold, temperatures and for a maximum period of between 10 and 14 days. If eggs are to be stored for the full 14 days, the temperature should be about 55°F (about 13°C). As storage time is reduced, the temperature can be raised slightly. For one week of storage, temperatures of 60–65°F (16–18°C) are adequate, while if eggs are stored for only two or three days, storage at room temperature, or 70–75°F (21–24°C) will suffice.

Left
Eggs ready for setting in a commercial incubator. Incubators provide a substitute for the hen by providing the proper environmental conditions of temperature, relative humidity, movement, and ventilation.

When exceeding the time factor for any temperature level, hatchability will decline at a relatively constant rate, with a warmer temperature hastening the loss of hatchability. If chicken eggs are stored at 55°F (13°C), virtually none will hatch after about five weeks.

Other related species respond to storage conditions differently. Of the common galliform-type production birds, ring-necked pheasant eggs will generally store for a maximum of about nine days, turkey eggs for about 20 days, and the champions, chukar partridge eggs, for 28 days before hatchability begins to decline.

Humidity during Storage

In addition to time and temperature, relative humidity of the storage environment is also important. Because of the porous nature of the shell, eggs constantly lose moisture to the environment. Eggs that lose too much moisture during storage and incubation have poor hatching success. To slow the water loss from eggs, high humidity—about 70 percent—during storage is necessary. The longer eggs are stored, the more important it is to ensure that the storage humidity is at the correct level.

Egg Position & Turning during Storage

During storage, eggs should be positioned with the blunt end up. Once again, it is not well understood why, but eggs stored blunt end down—or upside down—do not tend to hatch as well. Turning eggs during storage is unnecessary if they are to be set in seven days or less. If they are scheduled to remain in storage for longer than seven days, however, tilting them once or twice a day, from the first day of storage, will help to improve hatching success by 3–5 percent.

CHOOSING AN INCUBATOR

There is a wide variety of incubators available on the market for small producers. While they come in many shapes, sizes, and models, there are essentially two types: still-air and forced-draft incubators.

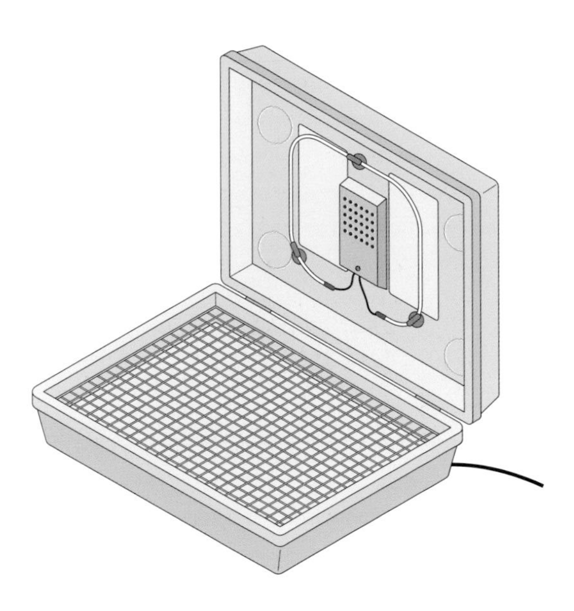

Still-air versus forced-draft incubator

A forced-air incubator uses a fan to circulate air around the incubator. This helps remove the risk of hot or cold spots that can occur with still-air systems.

Still-Air Incubators

Still-Air incubators are quite popular among small-scale poultry-keepers. They are generally tabletop models, holding from 3 to nearly 100 eggs. They are generally inexpensive, constructed of plastic or Styrofoam, and can incubate eggs in only one layer. They are available with upgrades such as automatic turning and conversion to a forced-draft type.

In all still-air incubators, air movement is accomplished by the natural process of convection, in which warm air rises and cold air sinks. The implication here is that there is a temperature gradient in the incubator, with air at the top warmer than that at the bottom, so eggs are kept in one layer only where the correct temperature is found.

Forced-Draft Incubators

Forced-Draft incubators contain a fan that moves the air, so the temperature is consistent in all areas of the incubator. With consistent temperature, eggs can be stacked throughout the incubator.

THE INCUBATION PROCESS

As with storing eggs before setting, the environmental conditions in which the eggs are kept during incubation, and the management techniques used, have a huge influence on hatching success.

Temperature during Incubation

Temperature is the primary environmental condition that controls the biological process of avian development. The increase in temperature that occurs when eggs are moved from storage to an incubator stimulates cell division, and embryonic development resumes. Note that it is important to have an accurate thermometer to measure the temperature inside your incubator precisely. Inexpensive card or dial thermometers are not accurate enough to check that the correct temperature for incubation is maintained.

Development is a very precise process, and cells and tissues must be formed at the proper rate to produce a normal offspring—21 days in chickens. In addition, temperature must remain constant, with only minor variations of less than 1°F (roughly 0.5°C), for proper development to occur. Research on artificial incubation over many years has determined the correct temperature for forced-draft incubation is 99.5–100°F (37.5–37.8°C), while still-air incubation should be at 102–103°F (38.9–39.4°C). Because of the virtual lack of air movement in still-air incubation, the temperature should be measured at the top of the eggs. Temperatures just 1–2°F (roughly 0.5–1°C) higher or lower will change the development rate, so that chicks will hatch either one or two days early or late, respectively. In addition, hatch success will be lower due to various potential abnormalities or other problems caused by improper temperature.

Humidity during Incubation

As is the case during storage, moisture continues to be lost from the egg throughout the incubation process, some of which is essential. On average during incubation, eggs will lose about 12 percent of their weight due to water loss. While eggs may hatch after a weight loss of anywhere between 8 percent and 24 percent, those that lose closer to the average of 12 percent will successfully hatch at a much higher rate. To control water loss, humidity in the incubation environment must be kept at 56–60 percent. Once again, when incubating outside of these humidity levels, eggs will either retain too much water during hatching, resulting in chicks that are edematous and "drown," or chicks that become dehydrated, dry out, and stick to the shell during hatching.

Many newer incubators have direct digital measurement of humidity, while older incubators usually use "wet-bulb" temperature. To accomplish this measurement, a moistened wick is attached to the bulb of a thermometer—due to evaporative cooling, the wet-bulb temperature will be lower than the dry-bulb temperature. For a humidity of about 56 percent, the wet-bulb temperature should be about 86°F (30°C).

Controlling humidity is accomplished by adding water to the incubator, usually in a small reservoir or pan placed in the bottom of the incubator. The humidity can be altered either by increasing or decreasing the surface area of the water reservoir, or by adjusting the ventilation to the incubator. Care must be taken if vents are closed to increase humidity not to allow the concentration of carbon dioxide to build up to lethal levels inside the incubator (see Ventilation during Incubation).

Ventilation during Incubation

During incubation, developing embryos require oxygen and release carbon dioxide in the same way as in hatched birds, only in much lower amounts, these amounts increasing as the embryos approach hatching. All incubators have vent holes of some sort or another that can usually be opened and closed for humidity control and to adjust the ventilation rate. The primary consideration here is to keep the carbon dioxide concentration inside the incubator below about 0.5 percent. Follow the incubator manufacturer's instructions on how to adjust ventilation, as they are all different.

Egg Position & Turning during Incubation

Egg position in most small incubators is horizontal, with the eggs laid on their sides. In larger incubators, eggs are generally set vertically with the blunt end up. Hatching success is generally about 5–7 percent better when eggs are set vertically, but horizontal setting is quite adequate.

Regardless of position, eggs need to be turned at least three times per day for proper hatching. If kept in the horizontal position, eggs are usually rolled to the other side, and if kept in the vertical position, they are generally tilted at a 45-degree angle, first to one side and then to the other. This daily movement prevents the growing embryo from sticking, thus increasing the hatching success. In addition, when eggs are set in the wrong position and/or not turned, the percentage of chicks that are in an incorrect position for hatching increases, again reducing hatching success. Turning should begin with the first day of incubation and continue through at least 14 days.

Candling

Although it is not required, candling eggs at around 7–10 days of incubation is good management practice. (Removing the eggs from the incubator for a few minutes as you do so will not disturb their development.) During candling, the egg is taken into a darkened room and a bright light is shone through it to observe the process of development.

Brown eggs are more difficult to candle due to their dark shells, so a brighter light and completely dark room are often necessary. White-shelled eggs are easy to candle. To make an inexpensive candler, cut a small hole (1 in. [2.5 cm] in diameter) in a piece of

Below

Egg candling is a way to observe the internal contents of an egg without breaking the shell. Candling is used to determine the developmental status of hatching eggs and the quality of eggs that are to be consumed.

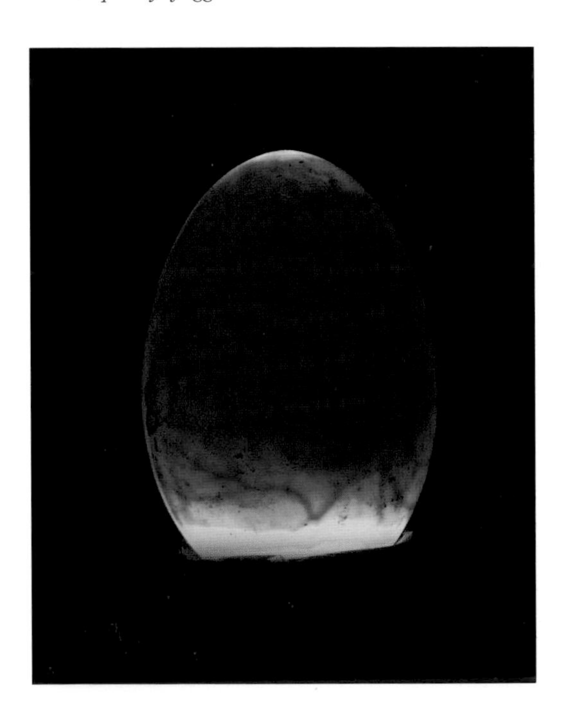

cardboard. Place a flashlight so that all the light is shining through the hole in the cardboard. Hold the blunt end of the egg against the hole so that all the light is aimed at the egg, which will be illuminated like a lightbulb.

In a normally developing egg at 7–10 days, candling will reveal a fairly large air cell at the blunt end, a somewhat dark area in the middle, and an obvious series of blood vessels. Movement of the embryo may also be noted. Following these quick and easy observations, the egg can be returned to the incubator so that the embryo can continue on its way to becoming a hatched chick.

Abnormal observations include clear eggs, which look like unincubated eggs. These are either infertile or the embryo died before completing the first day of incubation. A blood ring forms when an embryo has developed to the point of some blood formation (three to five days) but then died, resulting in pooling of the blood and giving the blood-ring appearance. A small dark spot (the dead embryo) may be noticed as well.

Occasionally an egg will appear to have dark fluid oozing from some of the pores, it may be slightly cracked, and it will emanate a noxious odor. If candled at this point, it will appear nearly completely dark. An egg with these characteristics is infected and has become home to one of the several bacteria strains associated with chicken eggs. These eggs should be immediately and carefully discarded. Occasionally an infected egg will explode in the incubator, or in your hand as you attempt to lift it from the incubator, spreading the infection to other eggs and requiring the complete and thorough cleaning of the incubator before the next setting.

HATCHING

For best hatching results, eggs should be transferred at about 18 days to a different incubator. Hatching is quite a dirty process, with the broken eggshells, chick down, and meconium (embryo feces) making cleaning difficult, so a dedicated "hatcher" is best.

At transfer, the eggs should be laid on their sides with enough space so they can move slightly. They will not roll around during hatching, and at this stage should no longer be turned. Turning at this stage can serve to disorient the chicks and reduce the possibility of hatching. The temperature should be reduced by about 1.5–2°F (about 1°C), and the relative humidity should be increased to about 65 percent or about 88–90°F (31–32°C) on a wet-bulb thermometer.

If the incubation conditions have been correct, the first chicks should pip (break the shell) and begin hatching at about 20½ days, with most out by the end of the 21st day—although there may be a few stragglers. It is good practice to allow the hatch to continue through day 22, after which any chicks that have not yet hatched probably will never do so. Even during good hatches, some will pip and not hatch, pip and die, or simply remain in the shell and not pip—this should be considered normal. You may want to "help" the chicks hatch, but this urge should be suppressed. Nature has made the hatching process difficult so that only the strong succeed. Late chicks often have physical or developmental problems that make them unable to hatch, and even if they are helped and removed from the shell, their chances of surviving the first week are poor. Any unhatched or partially hatched eggs can be humanely dispatched by placing them in a refrigerator for a few hours.

CHAPTER THREE

Introduction to Egg Quality

High-quality food is the desire of consumers worldwide, and purchasers of eggs are no different. Whether eggs are purchased from a local producer with 50 free-range hens or from retailers selling eggs from major egg producers with hundreds of thousands of hens, high quality is still paramount.

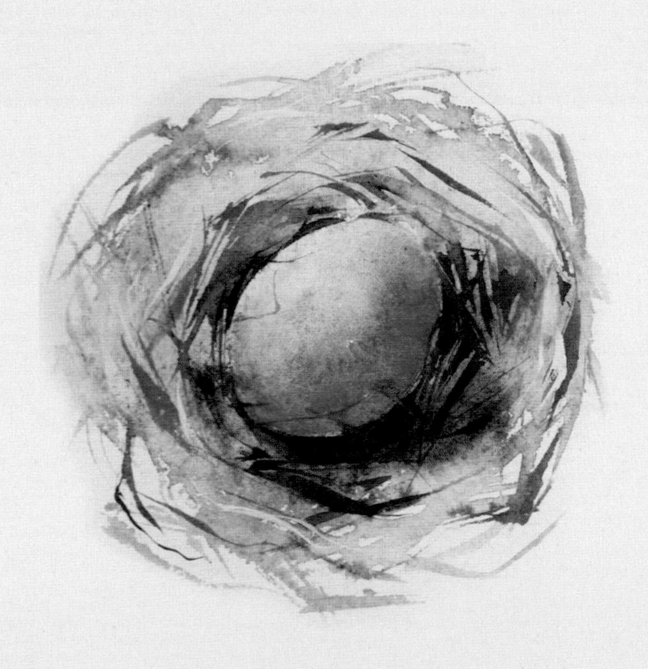

Quality Grading

The quality measure of eggs, unlike many other commodities, has little to do with wholesomeness or nutritional value and everything to do with appearance. Other measures—including shell or yolk color and egg size—while not technically related to quality measurements, are also important to consumers.

Each commodity has developed grading or quality nomenclature that is often specific to that commodity and not directly transferable. Quality grades, such as "Fancy" or "Extra Fancy," "Choice" or "Prime," and "Virgin" or "Extra Virgin," give consumers a general idea whether quality is high, medium, or low. Quality in eggs is based on their visual appeal, both at the level of the shell and the internal contents.

Before discussing egg quality, it must be noted that eggs are graded based on the lowest grade characteristic—the various grade components are not additive when determining overall grade. In other words, an egg with a high-grade shell and low-grade internal contents has an overall low grade, and an egg with several medium-grade traits of both shell and contents remains as a medium grade.

Egg-Quality Grading

	High	Medium	Low	Loss
US	AA	A	B	
Shell	Practically normal		Other	Cracked
Air cell	$<1/8$ in.	$1/8$–$3/16$ in.	$>3/16$ in.	N/A
Albumen	Clear and thick	Thin and watery, small meat spots		Large meat spots
Yolk	Diffuse yolk shadow		Blood spot $<1/8$ in.	Blood spot $>1/8$ in.
UK	Extra	A	B (broken/pasteurized)	Industrial
Shell	Sound and clean, unwashed		Other	Other
Air cell	≤4 mm	≤6 mm	>6 mm	Other
Albumen	Clear		Other	Other
Yolk	Diffuse shadow with no definite outline		Other	Other

External Egg Quality

Eggshells, which are not generally consumed, provide a single-serving natural packaging of eggs that makes food scientists green with envy. Most eggshells are of high quality as they are laid by the hen, but there are still several shell criteria used in grading.

SHELL SHAPE

Normal, high-quality eggshells are an ovoid shape, or a three-dimensional oval. This ovoid shape has only one axis of symmetry, the long axis, so when held vertically, the right and left sides are essentially the same shape. However, when examining the top and bottom halves of the ovoid, one is broad (blunt) while the other is narrow (pointed). High-quality eggs should have this shape. Lower-quality eggs have any shape other than this defined ovoid. Some eggs will be round, while others will be narrow and long, resembling a torpedo.

Other misshapen eggs include those with a wrinkled appearance, resembling a prune. During egg formation, following the addition of the shell membranes, the contents are typically "plumped" with metabolic water, which serves to distend the membrane to provide a smooth surface as a base for the shell. Sometimes the plumping is incomplete, so that when the shell is deposited, the membranes are somewhat flaccid and wrinkled. The shell simply takes on the shape of the membranes beneath with the typical characteristics of a hard shell.

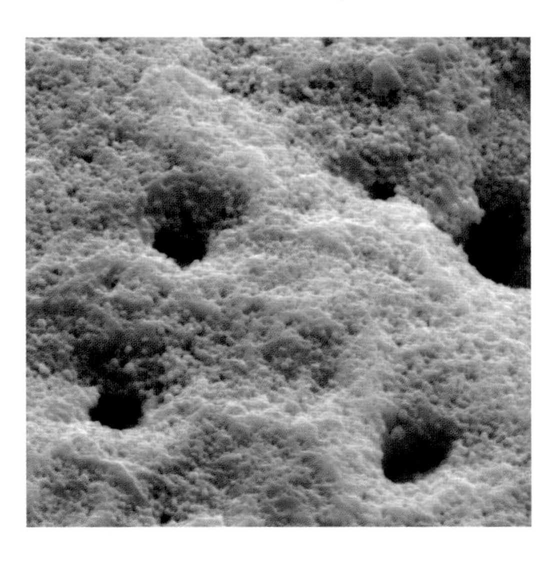

Left

The surface of an eggshell is covered with pores that are necessary for gas exchange and water loss during chick embryo development. These pores are also responsible for the water and carbon dioxide exchange of table eggs that contribute to the eventual loss of quality over time.

Defined ovoid of a high-quality egg

Torpedo shape of a low-quality egg

Round shape of a low-quality egg

Body-checked eggs are those that have the appearance of a bulge or "belt" around the middle of the egg, about halfway between the blunt and pointed ends. This occurs when the eggshell cracks in the oviduct of the hen before its formation is complete. Extra shell is deposited over the break to repair the damage, which is similar to how your own body repairs a broken bone. This region of thicker shell material gives the characteristic bulge of the body check.

Occasionally, two eggs end up in the oviduct at the same time. When this occurs, both eggs take on the characteristic shape of a slab-sided egg, or one with a flat side. Where the two eggs touch during shell formation, one side is flattened as the shell is deposited. The flattened side is often markedly thinner than rest of the shell, making it much weaker.

It is worth noting here that some diseases—often respiratory viruses—can cause misshapen eggs. Infectious bronchitis, Newcastle disease, and egg drop syndrome (not found in North America) are some of the important diseases in this group. In addition to the problems that they cause in eggs, clinical signs of respiratory disease will also be present, including noisy breathing (rales); coughing; drainage from the nares (nostrils), mouth, or eyes; and possibly a swollen face. These diseases are rare in small flocks of chickens, where most misshapen eggs are the result of other causes.

SHELL TEXTURE

High-quality eggshells should be relatively smooth to the touch. Shells with a rough, sandpaper-like feel to them are indicative of lower grades. In general, young hens produce smooth-shelled eggs, but as they reach 12 to 18 months in age, the shells of their eggs begin to have that sandpaper feel. Following the molt, the natural loss of feathers that usually occurs in the fall each year, shell texture improves again, losing some of the roughness. As post-molt production continues, the roughness of the shell increases more quickly and more severely, becoming quite obvious in the hen's second year.

Occasionally, eggs are produced with a sizable "blob" of extra calcium deposited in one or a few locations on the surface of the shell. These deposits can be from a few millimeters across to half an inch (1 cm) or more in size. Some have described the appearance as similar to a small amount of toothpaste squeezed onto the egg's surface. Determining whether a deposit constitutes a downgrade is somewhat subjective. If the consumer were to attempt to remove the deposit and this resulted in a hole in the shell, that egg would be considered poor in quality.

Why are shells sometimes rough?

The roughness is a consequence of the hen's reduced ability to form shells properly. Sometimes, increasing calcium in the diet can slow this process, but in the long run it is inevitable in most hens that lay into their third, fourth, and fifth years of life.

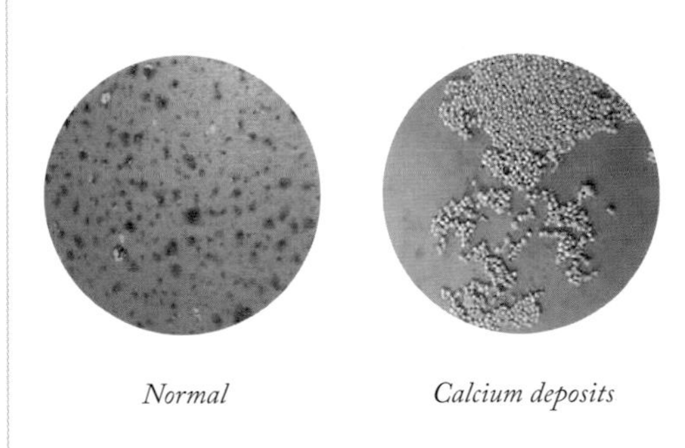

Normal *Calcium deposits*

SHELL CLEANLINESS

Clean shells are often the most obvious quality issue with eggs. Management of the hens, including regular cleaning of nests and coops, and hens "trained" to lay in nestboxes, is the best way to reduce the incidence of dirty or stained eggs.

Dirty eggs are those that have material adhering to them. This material can be litter from the nest, soil from the chicken yard, blood (generally from a small hen laying a large egg), or feces from the nest or feet of the hens. Typically, dirty eggs can be cleaned to remove this material, either by dry cleaning with a bit of sandpaper or by washing. (Note that care must be taken when washing eggs, as use of detergents, water temperature, time spent washing, and disinfection all can have negative consequences if done improperly.)

If, after cleaning, stains remain on the shell, they may or may not affect the quality depending on their size and prominence. The more visible the stain, the greater the possibility that the egg would be considered lower quality.

It should be quite apparent that when eggshell quality is considered, appearance of the egg is paramount. Consumers want to see eggs that look appealing. The United States Department of Agriculture (USDA), which regulates egg quality in the US, uses the phrase "practically normal" when referring to high-quality eggshell. Admittedly, "practically normal" is a subjective determination and is greatly dependent on the eye of the beholder.

Protection during processing, packaging, and transport are also of great importance to shell quality. Eggs with odd shapes and textures don't fit in the cartons properly, rendering them more likely to be cracked or completely broken between lay and consumer sales.

In the US, external egg quality grades are quite simple. Based on all of the quality issues noted above—shape, texture, and cleanliness—eggs receive a grade of either "AA" or "A" if their shells are "practically normal" and clean, or "B" if they have any shape or texture issues and only slight or moderate staining. Any egg that has material adhering to it is graded as "Dirty" and cannot be used for human consumption. The system is quite simple and somewhat easy to follow.

In the UK, the Department for Environment, Food and Rural Affairs (Defra) has a similarly straightforward system of egg-quality classification. Eggs with "normal, clean, undamaged" shells are Class A; those

with "normal, undamaged" shells are Class B; and those that are "cracked, misshapen, rough textured" or have "any other abnormality" are Class C. Eggs that don't reach Class C standards are classed as "Industrial Eggs" and are unfit for human consumption.

CRACKS

For all graded eggs, whether of high or low quality, the shells must be unbroken (sound). A cracked shell represents a major loss to the producer but a few of these eggs can be salvaged for use as human food by further processing. However, it must be noted that once the shell is cracked, the potential of contamination by pathogenic bacteria is exponentially increased, thereby increasing the possibility of foodborne illness. Cracked eggs from small flocks of chickens should be discarded.

Cracks are classed in two categories: the shell is cracked but the shell membranes remain intact and the contents are not leaking; or both the shell and membranes are broken and the contents are leaking.

Non-leaking cracks in commercially produced eggs, sometimes called "checks," are only used as a pasteurized liquid egg product. In general, along with many downgraded eggs with sound shells, these eggs are broken by machine and the contents (yolk and white) are separated and pasteurized. They are then sold in either large or small containers directly to consumers or to large egg users such as bakeries.

Leaking cracks are condemned and often processed into either animal feed or non-food industrial products.

Internal Egg Quality

The quality of the interior of an egg is based on a visual examination during candling, a method of shining a bright light through an egg to observe its contents (see pages 34–5). While the shell is solid, it is generally thin enough for light to penetrate sufficiently to determine enough information to assess the quality of its contents. White-shelled eggs are much easier to candle, whereas the contents of brown-shelled eggs are more difficult to see, especially if the light is dim and the room is light.

AIR CELL

The air cell of an egg typically forms at the blunt end after the egg is laid. During the egg's formation, its contents completely fill the interior. After it is laid, the egg begins to cool from the normal 106.5°F (41.4°C) deep body temperature of the hen to room temperature. While it is cooling, the contents contract within the ridged shell, which itself remains the same size. Because of the porous nature of the shell and the fact that there is a higher concentration of pores at the blunt end, the shrinkage causes a small space to form—the air cell. This air cell forms between the inner and outer shell membranes; biologically, it is used by the chick to take its first breath before hatching, but when considering table eggs, it is an indicator of quality.

Candling

When candling, there are three relatively obvious traits that can be seen which will indicate the quality of the egg: the air cell, the yolk, and the albumen. Each has visible attributes that can easily be observed to determine whether the egg is high or low quality.

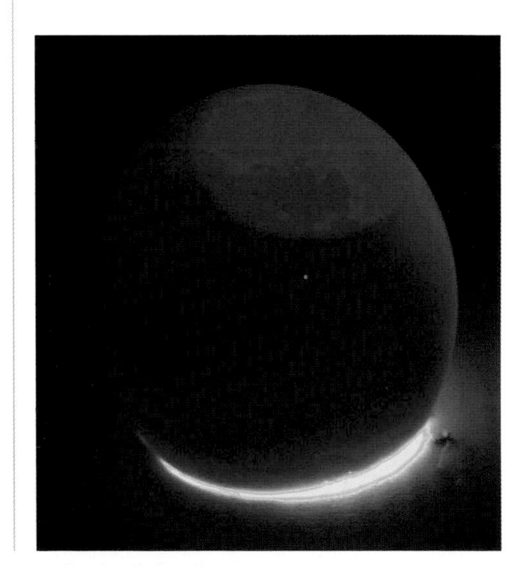

Right

The air cell increases in size as water is lost through pores in the shell. Smaller air cells are indicative of higher egg quality.

Immediately upon laying, the porous egg begins to lose water in the form of vapor. This water loss is fairly constant, as it is simple diffusion across the membranes and shell of the egg. As moisture is lost, air continually diffuses into the egg, taking up the space left by the lost water, and the air cell enlarges. In general, most eggs with high-quality shells will retain a small air cell for 30–45 days if stored properly. As the egg ages, the air cell continues to enlarge.

High-quality eggs have small air cells. In the US, the highest-grade eggs (Grade AA) must have an air cell that is no larger than 1⁄8 in. (3.2mm) when measured from the top of the blunt end. Medium-quality eggs (Grade A) have an air cell that is 1⁄8–3⁄16 in. (3.2–4.8 mm) deep. Any eggs with an air-cell depth greater than 3⁄16 in. (4.8 mm) are considered poor quality (Grade B).

Similarly, in the UK eggs with an air cell less than 1⁄4 in. (6 mm) deep are graded as Class A, and those with less than 3⁄16 in. (4 mm) deep as "Extra Fresh." Those with an air-cell depth less than 3⁄8 in. (9 mm) are Class B, and those with a depth greater than this are Class C.

Occasionally, the air cell will be misplaced so that it doesn't remain at the blunt end and instead floats around, or it may appear bubbly. Quality in these cases is estimated based on its size only, not on whether it is fixed or bubbly.

Testing for egg freshness

The size of the air cell and hence the freshness of an egg can also be determined by floating it in a bowl of water. If it floats it has a large air cell and tends to be old (it will have lost a considerable amount of water vapor, yielding a large air cell), while if it sinks it has a small air cell and is fresher. While shell thickness can play a role as to whether an egg will float or sink, this is a quick way to determine freshness.

Egg grades: air cells

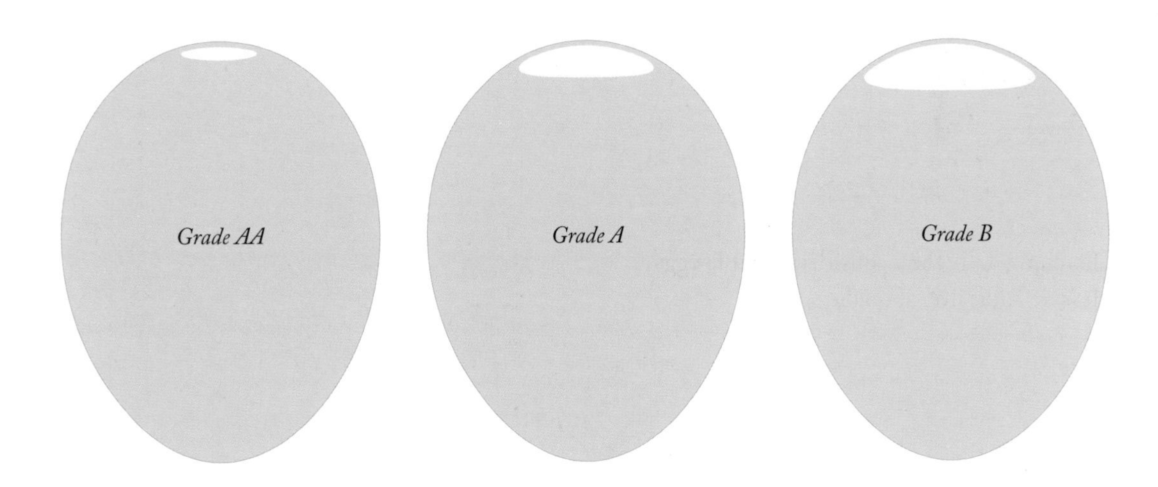

ALBUMEN

Upon candling, the albumen is observed for clarity and firmness. High-quality eggs have clear, firm (thick) albumen. During candling, the egg is rotated so that movement can be observed. Slow, viscous movement of the albumen indicates a high-quality egg, while more fluid, water-like movement indicates a poor-quality egg.

For many, it seems counterintuitive that the water loss causing the air cell to increase in size would also cause the albumen to become watery. To explain this phenomenon, a bit of chemistry is necessary.

At the same time as water is lost from the egg and the air cell increases in size, carbon dioxide that was dissolved in the albumen during egg formation is also lost from the egg. As the carbon dioxide diffuses from the egg, the pH of the albumen increases from a relatively neutral pH of about 7.5 to an alkaline pH greater than 9.0. Albumen is composed mostly of water (about 85 percent) and protein (about 12 percent). Each of the 13 different proteins that make up the albumen has a particular physical shape that is highly dependent on pH. As the pH increases, moving from neutral to alkaline, the protein structures change, reducing some of the tight bonds so that they become more flaccid, and hence the albumen becomes thin and watery.

Candling also allows any particulates in the albumen to be seen as shadows. Small pieces of ovary that break off during ovulation may end up in the albumen; these are called "meat spots." Small meat spots reduce the quality of an egg to a poor grade (Grade B in the US, and Class C in the UK), while those that are large and obvious reduce the quality to the point where the egg cannot be sold for human consumption.

YOLK

Observations of the yolk are dependent on the albumen. When the albumen is thick and firm, the yolk is held down in the middle of the egg, some distance from the shell. When the egg is candled, such a yolk appears as a dispersed shadow, more like a yellow or orange color distributed throughout the egg. As the albumen thins over time, the yolk will float closer to the shell, giving a more prominent shadow almost to the point that its outline can be visualized. The yolk is little affected by the process—only its appearance changes.

Blood spots on a yolk always result in the egg being graded as poor quality. These spots of blood are often wrongly considered as an indication of fertility. In reality, blood spots occur during ovulation. Occasionally, when the ovarian follicle opens to release the yolk, a small amount of bleeding occurs onto the surface of the yolk and is retained during the formation of the egg. Small blood spots (less than 1/8 in. or 3 mm) downgrade eggs to poor quality (Grade B in the US, and Class C in the UK), while anything bigger will render the egg unsuitable for human consumption.

Determining the shell quality and thickness

INTERNAL QUALITY BY BREAKOUT SAMPLING

As part of the regulation process in commercial egg production and in research, more accurate objective measurements of egg quality are often desirable. All of the previous determinations are quite subjective, relying on individual observers to provide a determination of quality, each of whom may have different requirements for grading. Objective measurements, however, give a more precise picture of quality.

Floating eggs in water can indicate their freshness (based on the size of the air cell). Eggs that sink in water have small air cells, which indicates freshness, while those that float have large air cells and should not be consumed.

Other objective measurements require breaking the egg, so only a small representative sample is used. Breaking the egg on a flat, level surface—often a piece of glass—provides the best way to observe and measure egg contents.

First, the shell's thickness can be measured directly following breakout. Usually several measurements are taken at the blunt end, middle, and pointed end of the shell, and the average is then taken. Normal chicken eggshell thickness is between 1⁄32 in. (0.7 mm) and 1⁄16 in. (1.5 mm).

The primary objective measurement of quality, however, is the thickness of the albumen. Using this and the overall weight of the egg, the Haugh unit of the egg is calculated, which is considered the most accurate measurement to determine egg quality.

Other observations include the overall appearance of the broken egg, such as whether the albumen spread is small or spread out, and whether the yolk stands tall or is flattened and may sometimes break. Both of these are dependent on the thickness of the albumen. Finally, any blood spots are noted during breakout observations.

Gravity on eggs

The specific gravity of eggs is one measure used to determine shell quality, or more precisely, shell thickness. Eggs are placed in various concentrations of salt solution, usually between 1.070 and 1.085 specific gravity. Eggs that sink at the higher densities have thicker shells and are of higher quality.

Storage of Eggs

The proper method of storing table eggs, from hen to consumer and beyond, is a controversial subject with public health professionals, producers, and consumers. Discussions concerning vaccinations of layers, washing of eggs, and refrigeration after packing have resulted in different requirements between countries. Some suggest that eggs can be stored at room temperature without any special packaging or temperature control, while others stipulate that refrigeration is absolutely necessary for food safety and to maintain quality. Washing and sanitizing eggs prior to packing to reduce the bacterial load on their surface is required by some and forbidden by others. Some countries vaccinate hens specifically to reduce colonization by *Salmonella* bacteria in the gut and ovary, while others have outlawed the practice.

In the early 1990s, following an outbreak of foodborne illness caused by *Salmonella enteritidis* contamination in the eastern US, eggs took on the status of a perishable food product, henceforth requiring refrigeration within hours of lay and throughout the market stream to the consumer. Labels on egg cartons strongly suggest that consumers also refrigerate eggs in the home. The cold temperatures found in fridges are known to slow the reproductive ability of bacteria, so if any food—including eggs—is contaminated, the number of bacteria remains few enough that the possibility of foodborne illness is greatly reduced.

In addition, there has also been a requirement in the US for many years that eggs are washed and sanitized under strict temperature and chemical controls to reduce the bacterial load. Producers of more than 3,000 hens are also required to test their environments for *Salmonella enteritidis*, with strict protocols of egg usage and subsequent cleaning put in place of bacterial contamination is found. More recently, vaccines have been voluntarily used by most large egg producers to reduce the incidence of *S. enteritidis*.

Egg safety has taken a completely different track in the UK, where neither washing nor refrigeration are allowed prior to sales of

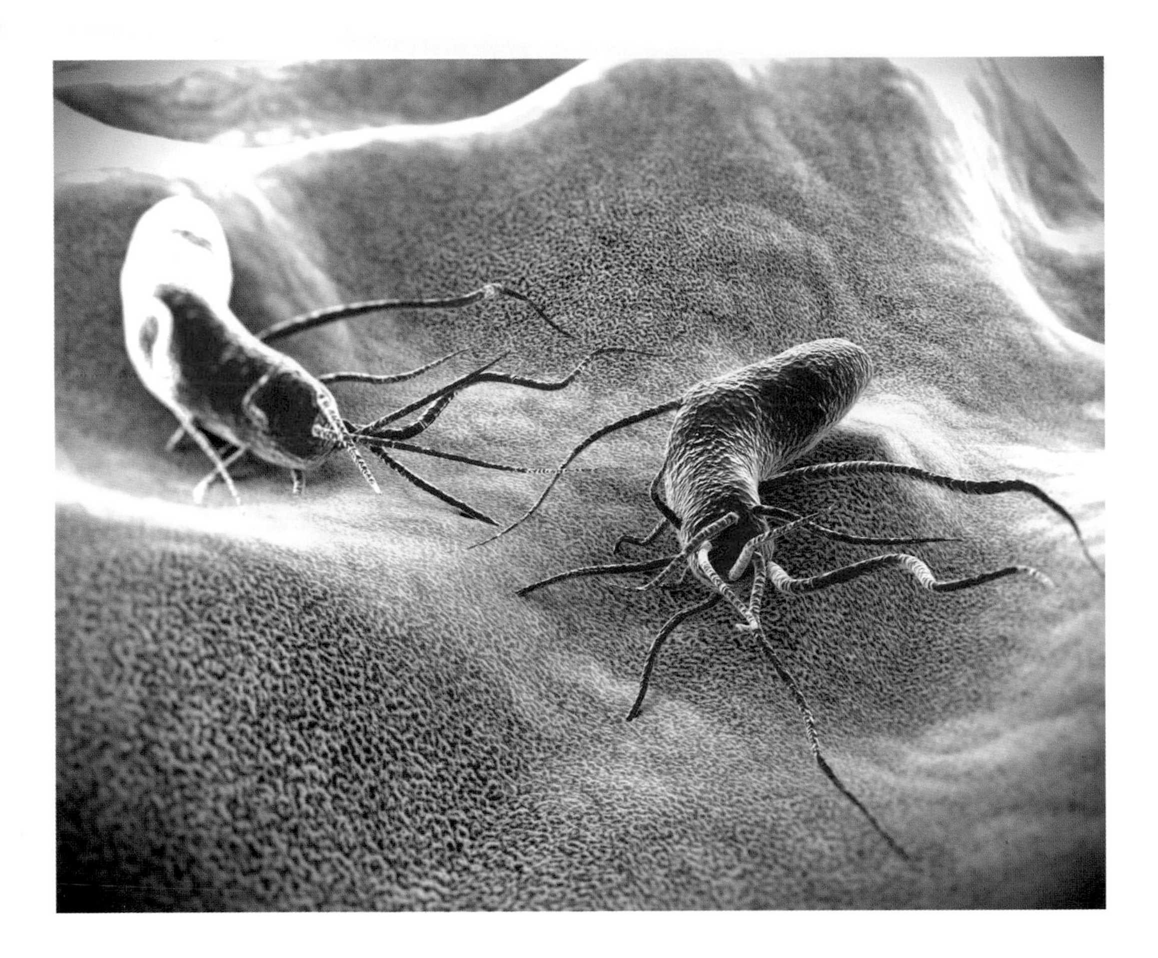

Class A eggs. Washing eggs removes their protective cuticle, which is known to reduce contamination of egg contents by surface bacteria, so in effect nature's defense against contamination is employed. Removing eggs from the cold temperatures in a fridge often results in water condensing on their surface, thereby increasing the possibility of contamination, which is why refrigeration is not recommended here either. Vaccination of hens for *S. enteritidis* is virtually required in the UK by most producers with some exceptions, and this has been quite effective in reducing foodborne illness caused by the bacterium in the UK.

Salmonella bacteria

An intact egg shell does not guarantee that it is free from bacteria because they can pass through the shell pores.

So, the debate continues; there is no consensus on the proper way to store eggs. Eggs will keep and maintain their quality longer when refrigerated, but if used quickly—within a week or so—the taste and baking characteristics of eggs at room temperature are better. Each producer and consumer should follow the laws and suggestions by their respective regulators. And as the final measure of food safety, consumers of eggs should cook them thoroughly prior to consumption.

CHAPTER FOUR

Feeding Chickens for Egg Production

Feed is an integral part of any production system, and raising birds in the best possible conditions is all for naught if the feed they consume does not provide the right nutrients at the right levels. While the general principles and ingredients are similar, feed formulations for different chickens do vary. This ensures that the birds receive the correct dietary building blocks in the correct quantities to maintain their health and, in the case of laying hens, to produce high-quality eggs in good numbers.

Commercial Feeds

Over the years, poultry nutritionists have identified all of the nutrients required by chickens of various ages, purposes, and levels of production. With this information at hand, they have formulated diets that are considered complete feeds, providing all of the requirements on a daily basis. These feeds are marketed to small-scale producers as sack feeds, which can be given to chickens as the sole source of feed throughout their lives. They are split into three types depending on the age or productive stage of the chicken: starter, grower/developer, and layer.

Overall, this feed-management system has been termed "phase feeding." With the system, chickens are not underfed or overfed, which improves their productivity in growth and egg production, and it addresses the cost of feed, thereby improving overall efficiency.

STARTER DIETS

Starter diets are fed to young pullets from hatch to about six weeks of age. Market chickens are often also fed chick starter throughout their growing period, from 6 to 10 weeks or so. However, chick starter is formulated for slow, sustained grow associated with laying chickens, not for the fast growth expected of a broiler. In some cases, starters that are specifically formulated for meat chickens and finishers, to be fed for the final two weeks or so before processing, can be purchased, which will support the much faster growth expected from a broiler chicken, hence increasing efficiency.

Generally speaking, starter diets contain the highest levels of crude protein, which ranges from 18 percent to 23 percent of the total diet for chickens. Starters with even higher protein levels (27–30 percent) are available for turkey poults and gamebird chicks. Starter feeds also tend to contain the highest levels of vitamins and trace minerals because of the higher requirements of these by fast-growing chicks, and the fact that the chicks can eat only small amounts of feed.

GROWER/DEVELOPER DIETS

These are fed to adolescent pullets aged from 6 to 18 weeks, just prior to point of lay. Most of the nutrient levels in grower/developer feeds tend to be either equal to, or lower than, those in starter diets. By the time these feeds are introduced, the growth rate of the birds has slowed and they are consuming enough so that the amount of each nutrient is adequate for their needs.

LAYER DIETS
Layer diets are fed to pullets and hens during egg production. They are lower in protein than starter and grower feeds but slightly higher in many of the other nutrients. While the bird's growth has essentially ceased, egg production requires higher levels of nutrients. Considering that the average hen lays about 20 dozen eggs each year, at an average weight of 24 oz. (680 g) per dozen, this works out as a total of 30 lb. (13.6 kg) of eggs, or at least six times the hen's body weight.

The usual constituents of layer feed include protein (17 percent), fiber (4.5 percent), fat (2.5 percent), phosphorus (0.5 percent), sodium chloride (0.4 percent), sodium (0.1 percent), and various vitamins and trace elements. The high levels of calcium (3.5–4 percent) form high-quality eggshells and maintain bone health.

ALL-PURPOSE FEEDS
Some feed suppliers produce "all-purpose" feeds that can be fed to chickens of all ages, from hatch through egg-laying. While feeding

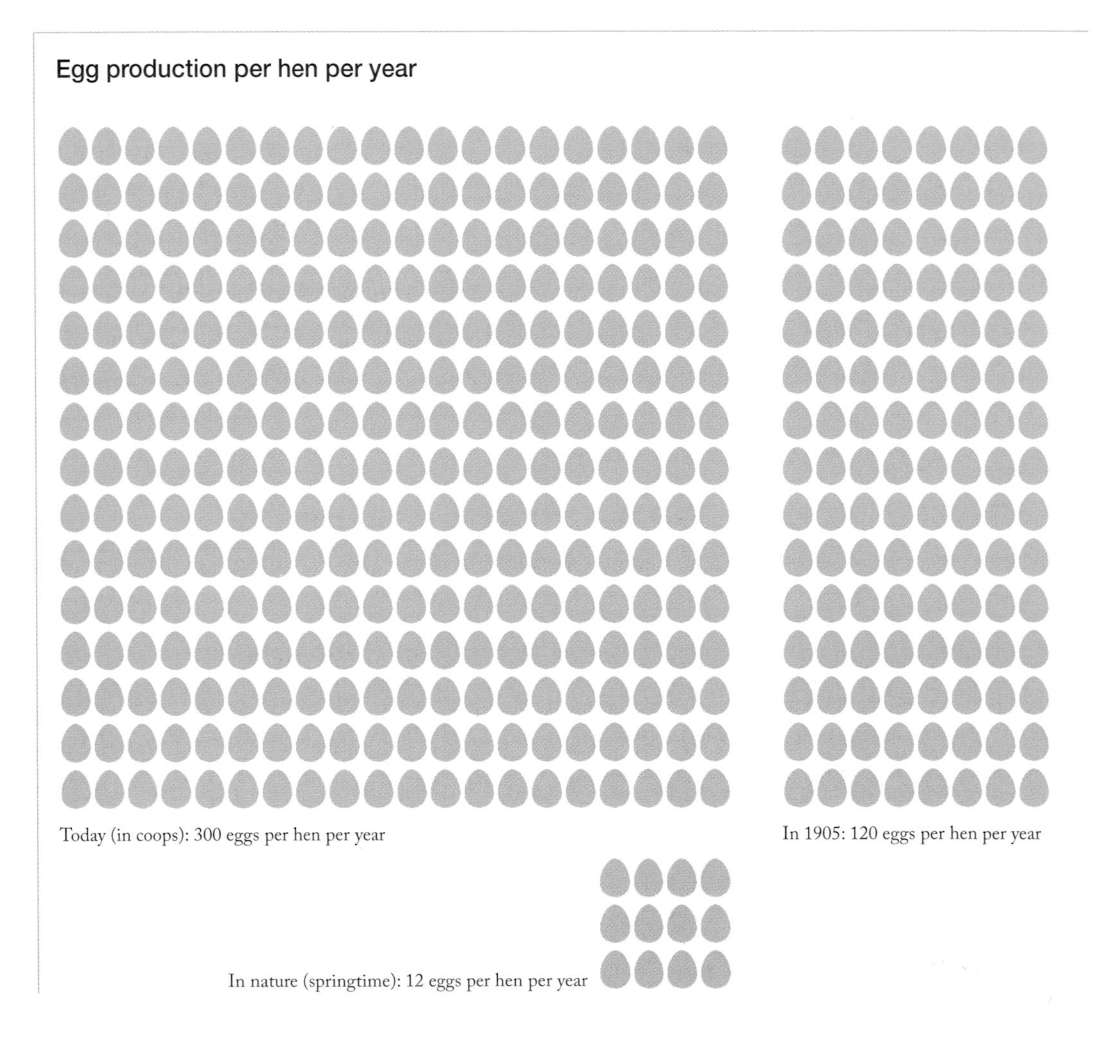

Feed grain constituents

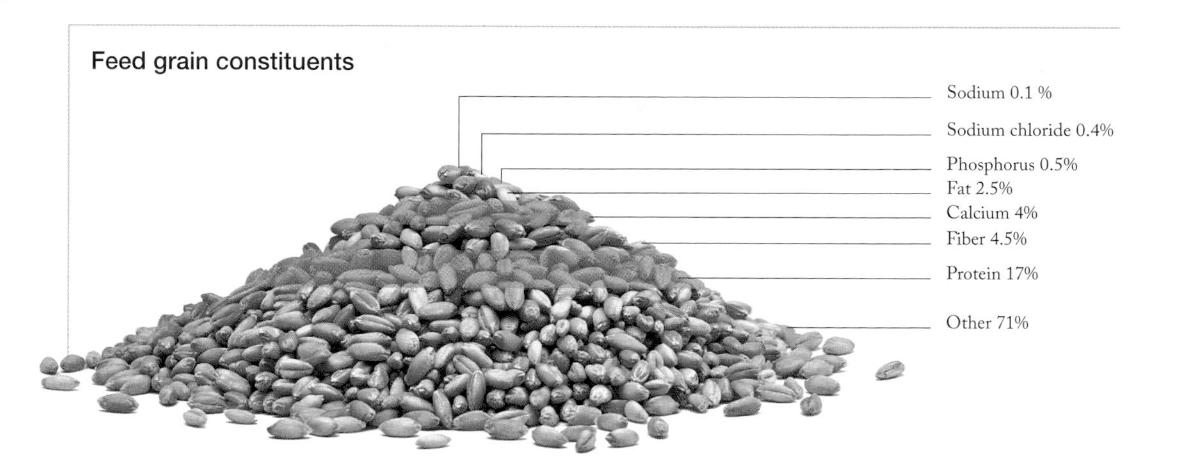

all-purpose feeds makes life easier, producers must realize that these feeds are not ideal for birds at any age. As such, supplements will be needed to ensure that the birds get the correct nutrients, including oyster shell as a calcium source for laying birds (see page 160). If you do choose to use an all-purpose feed, read the label carefully and provide any required supplementation as directed, or poor performance or worse could occur.

The need for supplements

Even with all the technology and variety of feed ingredients to choose from, not all nutrients can always be added naturally. The essential amino acid methionine, one of about 20 amino acids that make up protein molecules, is one such nutrient. The common ingredients used in poultry feeds tend to be low in methionine, which results in slow growth and poor performance. This problem became particularly noticeable when most animal products were removed from poultry diets. To counteract this potential deficiency, synthetic methionine is now added to most poultry diets.

NUTRIENTS VERSUS INGREDIENTS

When formulating diets, commercial feed suppliers use a variety of ingredients to supply the required level of nutrients. The nutrients are not added directly, but are included in various levels in the feed ingredients, which are then mixed in the proper ratios to provide the necessary levels. Common ingredients in poultry feeds include cereal grains for calories, including corn, barley, oats, and wheat; protein sources such as soy, canola, or animal products; and sources of other required nutrients such as limestone, dicalcium phosphate, fat, bonemeal, vitamin and mineral premixes, salt, and DL-methionine. The nutrient levels of each of the ingredients are known, so they can be mixed in the correct proportions to supply all the requirements for chickens at each stage of life.

During feed formulation, nutrient levels, availability, and price are all taken into consideration. All available ingredients and their current prices are entered into a computer, along with the calculated levels of nutrients for the particular feed being mixed. Based on price and availability, the formulation of ingredients

may change but the nutrient levels remain the same. This process is called "least cost feed formulation." Before the advent of this process, feeds were mixed much less accurately, with nutrient levels only closely approximating requirements. This resulted in lower efficiency through overfeeding or reduced productivity through underfeeding. Other supplements, including scratch feeds, oyster shell, and grit, are discussed on pages 160–5.

PROBLEM INGREDIENTS

"With every medicine there's a little poison" is a saying that is also true of feed ingredients. Along with their nutrients, many feed ingredients contain naturally occurring compounds that can have negative impacts on chickens. Most are in small amounts or are degraded during processing so they don't cause problems. However, some of these compounds can be quite toxic or interfere with digestion to the point that chickens don't perform well. These compounds include: protease inhibitors, which reduce the digestion of protein; lectins, which bind to epithelial tissue, causing damage; tannins, which bind with proteins and carbohydrates, making them resistant to enzymes; phytates, which bind phosphorus and other minerals, requiring extra phytase to counteract the effect; and non-starch polysaccharides, which resist enzymatic action, causing increased gut viscosity and fermentation.

Even common feed ingredients such as soy and wheat contain one or more of these anti-nutritional compounds. Feeding raw soy, which contains fairly high levels of protease inhibitors, reduces the effective digestion of protein, reducing growth and egg production. However, during processing, temperatures reach levels that denature these inhibitors, rendering them non-functional and resulting in an excellent feed ingredient.

Left
Foraging chickens are seeking seeds and insects. Their scratching behavior is a remainder from jungle fowl's scratching through the leaf litter on the jungle floor.

Wheat has high levels of xylan, a 5-carbon sugar that forms complex hemicellulose compounds that tend to hold a lot of water. When ingested at high levels (30 percent and above), wheat in the diet therefore results in poor digestion and wet droppings. In recent years, synthetic xylase, the enzyme needed to break down these compounds, has been added to feeds to aid digestion, making wheat more attractive as a feed ingredient.

Some potential feed ingredients contain problem compounds that cannot be "fixed," and so are not used in poultry diets. For example, dry beans such as pinto beans, kidney beans, and navy beans contain high levels of non-starch polysaccharides that can't be digested effectively, resulting in some fermentation and gas production. The growth of chickens fed 30 percent bean meal is reduced significantly compared with those not fed bean meal. Heating the ingredients or adding enzymes are ineffective in fixing the problem.

Cottonseed meal is another potential feed ingredient and is often fed to dairy cows as a protein source. When fed to laying hens, however, the compounds gossypol and cyclopropenoid fatty acids contained within cottonseed can cause green, brown, or mottled egg yolks and pink albumen to form in the eggs. Research to develop low-gossypol varieties of cottonseed have been successful, but the cyclopropenoid problem persists. Therefore, cottonseed is not fed to poultry. Similarly, alfalfa included in the diet of layers at levels as low as 3 percent significantly increases the number of eggs with blood spots.

Canola (rapeseed meal), which is a member of the mustard family of plants, has the tendency to cause a fishy odor and taste in eggs with brown shells. Canola contains high levels of the choline ester sinapine, which is quickly converted to trimethylamine (TMA), a fishy-smelling compound. TMA oxidase, the enzyme that breaks down TMA into an odorless compound that is then excreted, is synthesized at only low levels by brown-shelled egg producers related to the Rhode Island Red breed, including many Production Red strains of layers. Another compound in canola is glucosinolate, which further reduces the effectiveness of the TMA oxidase, allowing the TMA to be deposited into eggs and resulting in the fishy odor and taste.

In light of this, it is clearly important that producers seeking alternatives to traditional feed ingredients consider the potential problems when using unknown ingredients. While such ingredients may appear ideal, there are generally reasons why they are not used in commercial feeds.

FEED INGREDIENTS BENEFICIAL FOR EGG PRODUCTION

The list of beneficial feed ingredients is shorter than the list of problem ingredients, but they can be used quite successfully. Both brewing and distilling grains—the "mash" remaining after the production of beer or distilled spirits—are effective ingredients that have been shown to significantly increase the thickness of albumen when as little as 10 percent of the diet comprises these grains. These products require drying before they are mixed into poultry feed, or they will cause it to mold quickly.

Corn oil in the diet has been shown to increase both egg number and egg weight in young pullets during their first eight weeks of laying. And excess amounts of the vitamins choline and B12 in laying diets have also been shown to increase egg weight.

Organic Feeds

In recent years, many small producers have chosen to use organic feeds. Organic poultry and egg production is one of the fastest-growing commodities in the organic marketplace, and there has therefore been a great deal of attention on organic feeds for these producers.

There are different definitions of what constitutes organic production. In general, however, all feed ingredients used in organic production systems must come from certified organic farms that control the use of certain fertilizers and pesticides. More recently, the use of genetically modified (GM) ingredients has also been controlled in organic production. Finally, the addition of some synthetic products, specifically amino acids, are not allowed (with some exceptions) in organic feed formulation.

In general, there is little difference in the nutritional value of a feedstuff that is produced under organic or traditional methods, so diet formulations have little to do with whether the grain is organic or not. The problem arises when considering availability. Less certified

organic feed is grown, so these ingredients are harder to come by and consequently more expensive. Since the mid- to late 1990s, the proportion of GM corn and soy grown worldwide has increased steadily, to the point where more than 90 percent that is available can't be used in organic diets. While there are adequate non-GM substitutes for corn that are available for organic feed, the same is not true for soy. Virtually all substitutes for soy are significantly lower in protein and/or contain anti-nutritional compounds that negatively impact chicken growth or egg production. To add to the problem, the inability to add synthetic amino acids, especially methionine, at appropriate levels further restricts organic feed formulation. Consequently, organic diets are either quite expensive (two to three times the cost of non-organic feeds), or they use less expensive ingredients from foreign sources that may be of questionable nutritional value and quality.

Now that some of the science of chickens and eggs has been presented, the "art" of chicken and egg production can be better understood. As noted earlier, chickens and eggs have been a part of the human experience for millennia, with the science of poultry and egg production about 100 years old. With much research and a great deal of trial and error over many decades, practices for the efficient production of high-quality eggs and birds have been established. In addition, in more recent years, understanding of proper bird welfare and food safety has created systems of production that provide for the birds' well-being while at the same time producing high-quality, nutritious poultry and egg products.

Also note that while there are rules of thumb, chickens as a group are quite adaptable to various housing and management conditions. Therefore, as long as the basic necessities are provided, variation in methods can reflect any producer's individual needs.

Left
Cereal grains make up the bulk of feed ingredients in poultry diets. In the 1990s, genetic techniques were developed to place genes in common grains used in livestock feeds that impart resistance to diseases and herbicides.

2

BREEDS DEVELOPED FOR EGG-LAYING

Ancona

Features

The Ancona has black plumage with a white mottling on the tips of the feathers. The black should have a beetle-green tint and the white tip should be as V-shaped as possible. There are "diluted" blue and white variants of the breed, but these are infrequently seen and not recognized in the standards of perfection in all countries.

Origins

Debate surrounds the origins of the Ancona. It is without doubt an Italian breed that is named after that country's province of Ancona, but it came into being when sailors took white Leghorns (see page 70) with them on their journeys and then returned with other breeds from their trading destinations. There are no records of what was mated with the original Leghorns to create the Ancona, but it is fair to speculate that it was probably a breed that is now extinct.

Hen

The Ancona is a light breed, the hen weighing around 4.5 lb. (2 kg). The body is broad to the eye with a full tail that is carried high. The head is carried well back and the comb is either rose or single. In the latter case, the comb should hang elegantly over the face without obstructing vision.

The Ancona exhibits a marked resemblance to another well-known layer of large quantities of white eggs: the Leghorn. It is, however, viewed as a breed in its own right and has its own dedicated following of owners.

Cock
The Ancona cock is a very typical country fowl or light breed. He weighs around 6 lb. (2.7 kg) and has an alert, active, and gallant appearance. As with the hen, he can have either a rose or single comb, but in the case of the latter it should be upright, of a medium size, and with five to seven serrations.

Temperament
As with all types of country fowl, Ancona chickens are constantly busy throughout daylight hours and as such it can take some effort to tame them. Given the correct treatment, though, by a patient keeper, it is possible to gain their trust.

Care
Developed for their egg-laying capability and enhanced foraging skills, Anconas are highly capable of flight. They are best suited to a free-range environment, a husbandry system that will see them flourish. They are very tolerant and fit well within most climates, needing little in terms of additional or special care.

Egg
The egg of the Ancona is white, and hens will lay a considerable number in a year. They competed in many laying trials during the early part of the 20th century, garnering a strong following with egg producers in some regions. They are not prone to broodiness.

Araucana

Features

The Araucana's face muffling is not seen on birds in North America. Another notable difference in North American Araucanas is the lack of last vertebrae, giving a "tailless" or "rumpless" appearance. Depending on which country's standards of perfection are used as a reference, the precise characteristics of the breed can vary significantly. In terms of plumage varieties it is most frequently seen in lavender, but a large range of other colors is accepted.

Origins

The Araucana originates from Chile in South America, and it is widely held that the name is derived from the Araucanía territory in the north of the country. The indigenous Indians bred this form of jungle fowl and it was later obtained by the Spanish when they traveled to South America. Little is known about its ancestry, and scientists today are still trying to establish if, in its original form, it is linked to the ancestors of the domestic chicken or is in fact completely unrelated.

Hen

In North America, the Araucana hen weighs about 4 lb. (about 1.8 kg) and is slightly larger in the rest of the world. The illustration here shows the abundant face muffling for which the breed is known in most countries; in North America, breed has ear tufts but no muff on the face or throat. The hens are infrequent broodies but, should they sit, they make very good mothers.

A majestic and striking breed, the Araucana is famed for the exceptional range of beautiful blue eggs that it lays, along with its distinct appearance. The images here illustrate the type of feathering that is seen on the breed everywhere in the world except in North America.

Cock

The Araucana cock has an alert and active carriage with a long, deep body that is firm to the hand. Weighing around 5 lb. (2.3 kg) in North America and slightly heavier elsewhere, it is an average-sized bird with large wings. North American males lack the long tail feathers shown in the illustration.

Temperament

Araucanas are fairly placid birds that can be tamed easily, soon learning to trust their keeper.

Care

The Araucana is a hardy breed that exhibits a high level of vigor. The birds cope well with poor weather conditions and are suited to both free-range and more confined husbandry setups. They are capable of short flight but are not prone to taking to the wing, so medium-height fencing will be sufficient to contain them. The crest size can vary according to the country's standard of perfection, but care should be taken not to allow it to cause vision impairment, and it should be checked regularly for the presence of lice.

Egg

The Araucana egg is one of the most striking in the chicken world. The color can vary through a range of blues, from light pastel to almost olive green. The other unique attribute of the eggs is that the color permeates throughout the shell, inside and out.

Australorp

Origins

The Australorp was developed prior to 1920 in Australia, when some of the first black Orpingtons produced by William Cook in Kent, UK (see page 78), were taken to Australia. There they were developed further into highly efficient laying birds with good utility qualities. The resulting Australorp breed became an attractive proposition for poultry-keepers in other countries, and it was first exported in the early 1920s.

Features

The Australorp can be seen in white, blue, and black plumage types in its native land, but the best known color, and the only color recognized in North America, is black. It has a similar outline to its ancestor, the Orpington, but its feathers are tighter to the body, its shape is marginally longer, and its stance is more erect. The breed has also maintained more of its utility qualities than its British predecessor.

Hen

The Australorp hen is a graceful, well-balanced bird weighing 6–8 lb. (2.7–3.6 kg). With a deep, broad body and sweeping curves, she should appear longer in length than in depth of frame.

The Australorp is probably the most widely distributed breed of chicken with Australian roots. As the name suggests, it is an abbreviation of "Australian Orpington," although careful breeding has developed subtly different characteristics from its UK progenitor.

Cock

The posture of the Australorp cock denotes his active behavior. Weighing 8–10 lb. (3.6–4.5 kg), he is alert, erect, and flowing, and is clean-legged and free from any coarseness in appearance. It is not surprising that "type" is the key point-scorer on the show bench for this breed.

Temperament

Despite being lively and active, the Australorp is a friendly and mild-mannered breed to keep. The birds are tolerant toward other breeds and, in fact, young males show little if any aggression to each other, enabling the keeper to pen them together. Easy to tame and engaging to keep, they make a good starter breed for beginners, providing the space needed to accommodate this larger layer is available.

Care

Australorps are not keen flyers, so relatively low fencing around 5ft (1.5m) high will be sufficient to contain them should a fixed run be the husbandry option used. They do, however, benefit from a more free-range existence, as they are quite capable foragers. Unlike their cousins, the Orpingtons, the hens rarely sit, so broodiness is seldom an issue.

Egg

The Australorp egg falls in the range of tinted to brown. The hens make great layers—they succeeded in many laying trials during the 1930s and 1940s and won a number of major shows at the time.

Barnevelder

Features

The Barnevelder has been seen in a range of plumage types over the years, but it is probably best known for the intricate double-laced variety. Other plumage types commonly seen are the black, partridge, and silver. A well-balanced breed, the Barnevelder has tight feathering, bright, prominent eyes, and a medium-sized single comb and wattles.

Origins

The Barnevelder was developed in the town of Barneveld in the Netherlands just prior to the onset of World War I, putting it among some of the most recent pure breeds to be created. It is believed that Cochins, Brahmas, Langshans, and Malays constitute part of the Barnevelder's makeup, along with various undocumented crossbreeds.

Hen

The Barnevelder hen is a deep-bodied, medium-length bird with a high-set saddle. She weighs 6–7 lb. (2.7–3.2 kg) and, while she can sit, she is not noted for her broodiness.

The Barnevelder is probably best known on two counts: first, for its beautiful mahogany and black double-laced plumage, which is by far the most popular type seen; and second, for its ability to lay dark brown eggs whose color can surpass that of the eggs laid by its countryman, the Welsummer (*see page 84*), and contend with its neighbors, the Marans of France (*see page 72*).

Cock

The Barnevelder cock weighs 7–8 lb. (3.2–3.6 kg) and has an alert, upright carriage with a graceful tail that sweeps into a broad body. Both sexes and all colors have yellow legs.

Temperament

The Barnevelder has a placid nature, and if handled carefully by the keeper will become tame and trusting. The birds are tolerant toward each other and will also exist happily within a mixed flock of breeds.

Care

As an average-sized breed, the Barnevelder is capable of getting off the ground but is not prone to flight, so a medium-height fence will keep the birds contained within your property. The main downside of this good layer breed is its susceptibility to Marek's disease (see page 168). Buyers should consequently look to purchase vaccinated stock to avoid disappointment, and breeders who are serious about rearing Barnevelders should give careful consideration to a vaccination program.

Egg

The Barnevelder egg is a dark brown color comparable with that of Marans eggs. Coupling this feature with the sophisticated plumage makes it an attractive breed to keep. Its egg color does, however, fade quickly as the hens' laying season progresses, so don't expect consistency throughout the year.

Lakenvelder

Origins

There is debate about the precise origins of the Lakenvelder. Some experts place it as having German roots, whereas others have traced it to Holland as far back as the 1720s. The former believe the name to originate from the fact its coloring resembles the Lakenvelder cow whereas the latter claim that its name comes from Utrecht hamlet of Lakenvelt in the Netherlands.

Features

The Lakenvelder is a very attractive-looking bird due to the boldness of its coloring, making it stand out against the surrounding countryside. It is very active, moving around constantly as it forages, and is quite capable of flight, reaching the tops of trees with ease. Although it comes in only one plumage type, it can be very difficult to breed good examples. However, individuals that do meet the mark are extremely handsome birds.

Hen

The Lakenvelder hen weighs 4.5 lb. (2 kg). She carries a neat, compact single comb , which has the unique characteristic of standing upright in the female, and delicate wattles. Her appearance is spritely and her body is long, with the tail carried at a 45-degree angle.

The Lakenvelder (also known as the Lakenfelder) is a striking breed that is both an excellent layer and has very reasonable table qualities. The Lakenvelder has Threatened status with the American Livestock Conservancy. Its belted markings give it a very distinct appearance.

Cock

The Lakenvelder cock has a bold, upright posture. Like the hen, his character is spritely and, allowing for the usual differences in feather form between the sexes, he exhibits a similar plumage pattern. He weighs 5 lb. (2.3 kg).

Temperament

As is the case with the lighter country-fowl breeds, the Lakenvelder is not easily tamed, preferring instead to keep its distance from even the most patient of keepers. As such, it is not really suitable as a beginner's breed.

Care

Lakenvelders are most definitely better suited to free-range setups—in fact, their spritely, spirited character can lead to them being quite jittery if kept in a confined space. This in turn can impact on their productivity as a laying breed. They are fast-growing and vigorous, and cope well in all weathers. No special care measures are required other than to ensure that run systems have high fencing or are roofed in to contain these competent flyers.

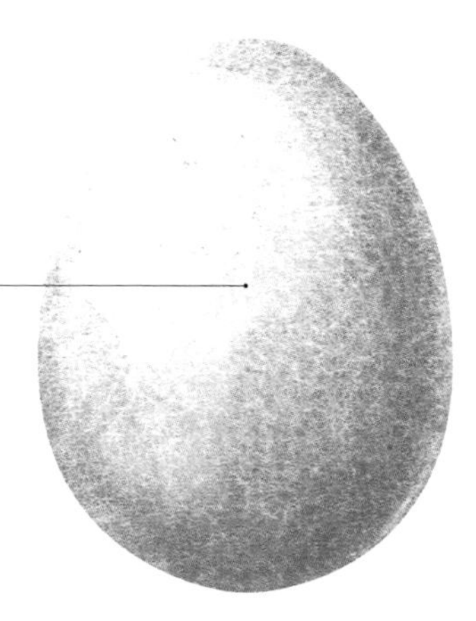

Egg

The egg of the Lakenvelder sits in the white to tinted range. The hens are enthusiastic layers and are typically non-sitters, so broodiness is seldom a problem.

Legbar

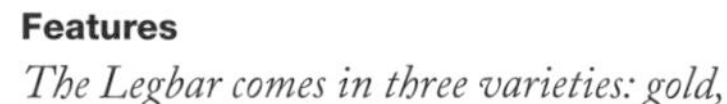

Features
The Legbar comes in three varieties: gold, silver, and—probably the most well known today—cream. The latter has a small crest on its head carried well back from the eyes. The comb is large, single, and serrated, and is coupled with thin wattles. The body is wedge-shaped with a long, flat back that slopes slightly toward the tail. The Legbar is an alert breed that lays as well as its constituent breeds.

Origins
The Legbar was created in Cambridge, UK, and is accredited to the breeding work of Reginald Punnett (deviser of the Punnett square) and Michael Pease in the 1930s. They uncovered the basic principles of the barring gene in their experimental work on chicken breeding when trying to establish a mechanism to sex chicks accurately based on plumage appearance. The Legbar is not currently recognized by the American Poultry Association.

Hen
The Legbar hen weighs 4.5 lb. (2 kg). She carries her wings tight to the body and her tail at a 45-degree angle to the back. The comb can be upright or fall gracefully to either side of the face, but not so much so as to obstruct eyesight. The crest of the cream Legbar hen tends to be fuller than that of her male counterpart.

The Legbar is an autosexing breed, meaning chicks can be sexed at hatching. The chicks are separated on the basis of their plumage, which differs between males and females thanks to the barring gene inherited from their barred Plymouth Rock (*see page 80*) parentage. The Legbar's other parent breed is the Leghorn (*see page 70*), while the cream Legbar is the same cross but with an added "dose" of Araucana (*see page 60*) to provide the head crest and interesting egg colors.

Cock

The Legbar cock has an upright and alert carriage. He weighs in at 6 lb. (2.7 kg), which reflects the Legbar's status as a light breed. The comb is upright and evenly serrated in each of the gold, silver, and cream varieties, but the cream Legbar cock also carries a crest of feathers on his head.

Care

Legbars are hardy birds that are well suited to a free-range existence, as they are excellent foragers. They cope well in most climates and require little specialist help. They are, however, capable of short flight and can reach a reasonable height, so roofing is required if they are to be contained within a run system. The cream Legbar crest does need a little more attention—as with all crests, it can harbor lice and mites more readily.

Egg

The eggs of the gold and silver Legbars are white or cream in color. The eggs of the cream Legbar range through blues, greens, and olives due to the presence of Araucana genes within its makeup.

Temperament

Legbars can be quite flighty, taking on a lot of the sprightly characteristics of their Leghorn parentage. It is possible for a patient and calm keeper to hand-tame them, but the birds tend to maintain their distance in most cases.

Leghorn

Features

The Leghorn is very much a lightweight breed, with early examples weighing as little as 3.5 lb. (1.5 kg). Crossing Leghorns with Malay and Minorca birds (see page 74) has increased their weight, but their spritely and alert appearance—coupled with a long back and lengthy, tilted tail—means they remain lean-looking. There are 16 accepted colors in the American standard, and while exhibition breeding has had an impact on productivity, many strains are quite capable of laying in excess of 200 eggs per year.

Origins

The Leghorn is a Mediterranean breed that originates from Italy, although it was further developed by a number of countries during the late 19th century, most notably the US. It was exported from there to the UK in 1870. Its wide distribution has led to a number of standards of perfection being developed, within which there are some marked differences buyers should be aware of when purchasing stock.

Hen

The Leghorn hen should weigh around 4.5 lb. (2 kg). Her feathering is tight to her body and her tail is carried at a slight angle. She comes with either a rose or single comb, which in the latter case should fall elegantly over the side of the head but not so much so as to obstruct the hen's vision.

The Leghorn has contributed significantly toward the development of commercial and hybrid egg-laying breeds over the last 60 years. Its capacity to produce high volumes of eggs from such a relatively light frame, together with its instinctive ability to forage for food, thus keeping feed costs down, make it a very economical breed to own.

Cock

The Leghorn cock is an attentive and active-looking bird. Unlike in the female, the single comb is large and stands erect, and is joined by long, thin wattles hanging from the underside of the head. The ideal weight is around 6 lb. (2.7 kg).

Temperament

The Leghorn is not known for taming easily. They are fast, flighty birds and cautious of any keeper, but by the same measure they are very curious and make inquisitive members of any flock.

Care

This is a robust breed suitable for almost any weather conditions, although additional care should be taken to protect the large comb and wattle in freezing conditions. The Leghorn is an excellent forager and hence best suited to a free-range lifestyle. The birds are quite capable of short flight, so high fencing or roofing is required if they are to be contained within a run area.

Egg

The Leghorn egg is white. The hens are prolific layers, generally producing an egg a day without a break when in lay. They rarely become broody, but if they do, they will be fiercely protective of their young.

Marans

Origins

The Marans takes its name from the town of the same name on the midwestern coast of France. Its ancestry lies with a number of breeds, including Faverolles, Langshans, Coucou de Malines, and Plymouth Rock.

Features

In their native country and in the US, Marans with feathered legs are favored, whereas in other countries, such as the UK, featherless legs are selected for within the breed. Depending on which country's standards of perfection are used as reference, various plumage types can be seen, with black, dark cuckoo, silver cuckoo, and golden cuckoo being the more common.

Hen

The Marans hen weighs 6.5 lb. (2.9 kg). As with the male, she has white legs, red or bright orange eyes, and a medium-sized white or horn-colored beak. She has a compact stance with neat, medium-sized wattles and a single comb.

The Marans is probably the best-known breed of chicken originating from France, primarily because of its ability to lay incredibly dark brown eggs. It is also suitable as a table bird and is widely used for both meat and eggs in some regions. It is, however, for its egg color that many breeders keep the Marans, and today the birds are not as widely exhibited for their looks as for their eggs.

Cock

The Marans cock has an active-looking posture. Weighing 8 lb. (3.6 kg), he has a medium-sized, solid-looking body, a strong neck, and a relatively long, upright tail. With the reasonable width and depth of frame of the male, it is possible to see the underlying table qualities of the breed.

Temperament

As a breed, Marans chickens exhibit a high level of curiosity and will actively seek out their keeper. They do, however, rarely become tame, preferring instead to remain at arm's length. They are friendly among themselves and with most other breeds, seldom showing aggression.

Care

Being effective foragers, Marans are well suited to a free-range environment, although they will also perform well in a more confined run. They are a vigorous breed capable of coping with any climate. The feathering on the legs of the French and American strains of the breed is less profuse than in most other feather-legged breeds, so they cope sufficiently well in wet or muddy conditions without need for special care.

Egg

The Marans egg is extremely dark brown, and its actual color quality can be assessed using an agreed and accepted points scale. The pores on the egg surface are smaller than in eggs of other breeds and the membrane can be very thick, particularly in darker eggs. This can lead to hatching problems.

Minorca

Origins

The Minorca is a development of the common black fowl of Spain known as the Castilian. Its name is believed to have been derived from the port of Ciutadella de Menorca on the western end of the Balearic island of Minorca, from where many of the birds were exported to other European countries such as the UK and Germany.

Features

The Minorca is one of the heaviest of the light breeds and is popular on the show bench. Although it is predominantly seen with black plumage, white and buff colors are also seen, and there is even a blue version in the UK.. The main characteristics of exaggerated lobes, combs, and wattles remain significant in the remarkable appearance of the bird.

Hen

The Minorca hen weighs 7.5 lb. (3.4 kg), and has an upright carriage and a graceful appearance. This is accentuated by the lop of the single comb, which drops down over one side of the face but not so much as to obscure the bird's vision.

The Minorca is a striking-looking Mediterranean breed of chicken that is well known for its prominent white earlobes, surpassed perhaps only by those of the White-Faced Black Spanish breed. It is a typical-looking southern European laying bird, with a dedicated following on the exhibition circuit.

Egg

The Minorca is famous, and highly prized, for its ability to produce extra-large white eggs. However, the hens seldom go broody, and therefore eggs need to be hatched by another breed or in an incubator.

Temperament

As with most light breeds, the Minorca is capable of short flight and can easily fly quite high. The birds can be a little skittish, but a keeper who remains relaxed and patient with them will be rewarded with a friendly flock.

Care

The Minorca is a hardy breed that can cope with most climates, but care must be taken with the bird's headgear during the winter months. The exposed skin of the large lobes, wattles, and combs should be treated with petroleum jelly to protect them from frostbite, and drinkers need to be adapted so that these features stay dry. As the Minorca is a light breed capable of flight, a roof will be required if the birds are contained within a run. If they are free-range, then suitable boundary fencing may be needed.

Cock

The Minorca cock weighs up to 9 lb. (4.8 kg) and shares the same alert and graceful appearance as exhibited by the hen. However, the single comb stands upright and has six even points. Rose combs are acceptable as well.

New Hampshire

Origins

The New Hampshire was developed in the 1930s in the state of New Hampshire using predominantly Rhode Island Red birds from the neighboring state (see page 82). It is believed that Wyandottes (see page 86) also played a small part in its development. The New Hampshire was admitted to the American Standards of Perfection in 1935, making it one of the most recent pure breeds to be recognized. Although popular in the United States, it is frequently considered a rare breed elsewhere.

Features

The New Hampshire shares a lot of the qualities of its neighbor, the Rhode Island Red. It is a fast-maturing, quick-feathering, weighty breed that will lay a good number of eggs. The plumage ranges from chestnut red to reddish bay, with beetle-green-tinted black coloring in the tail feathers. In some countries the tail can also be seen in an attractive blue color.

Hen

The New Hampshire hen has a fairly consistent shade of chestnut red throughout her plumage, with the lower neck feathers tipped in black. With a neat five-point single comb, medium-sized wattles, and a broad, well-rounded body, she ideally weighs around 6.5 lb. (3 kg).

As its name suggests, the New Hampshire is closely related to the Rhode Island Red (*see page 82*), not just in terms of geography but also in genetics. It's a great laying breed, which, like its neighbor, can also deliver good table qualities.

Cock
The New Hampshire cock weighs 8.5 lb. (3.9 kg), and is alert in posture and well balanced in appearance. His plumage ranges from a brilliant reddish bay on the head to a deep chestnut red on the wings and saddle, grading to black within the tail.

Temperament
The New Hampshire is a friendly and tolerant breed. The cocks are not known for aggression, and both sexes can easily be tamed. Broodiness does occur in the hens, although it is not a strong behavioral trait in the breed.

Care
New Hampshireare vigorous, healthy-looking birds that are well suited to either a free-range existence or being confined within a run. They are heavy and quite stocky, so are not good flyers and as such can be secured behind a 5ft-high (1.5m) fence. Adaptable to many climates, they are capable of handling most weather conditions with ease.

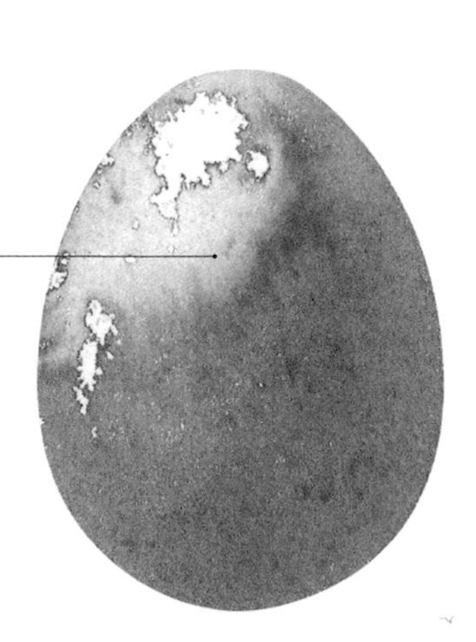

Egg
As is true of its neighbor, the New Hampshire lays a good number of tinted to brown eggs.

Orpington

Origins

The origins of the Orpington are well documented, the breed having been developed in the village of Orpington in Kent in the UK. Its creator, William Cook, wanted a breed of poultry that could produce large clutches of brown eggs during the winter months and ample meat from a plump breast. He also wanted a breed of fowl that was "of handsome appearance."

Features

The Orpington is a large, broad-looking bird, its immense size being further enhanced by an abundant and profuse mass of feathers. Its bulky appearance is instantly recognizable, such that the shape and outline of the bird plays a significant part in the standard of perfection for the breed. The breed is seen in black, white, buff, and blue color varieties, with buff being the most popular.' The broad feathers should fit smoothly to the body.

Hen

The Orpington hen is a graceful bird with a short back that gives way to a wide but almost flat cushion and a compact tail.

The Orpington has an impressive pedigree and is well known as a majestic breed. Originally, however, its purpose was slightly different, with much more focus on the livestock values of egg production. In its day, it served both functions very well, in particular as a winter egg-layer.

Cock

The large, bold eyes of the Orpington and its neat, medium-sized comb and wattles give the breed a distinctive face. Standard weight for the cock is 10 lb. (4.5 kg).

Care

Due to their large size, Orpingtons do require suitably proportioned housing. They are equally comfortable living in a fixed run as they are with free-range setups. They will cope in most climates, but as they can easily get soaked in wet weather, it is important to provide them with sheltered areas.

Temperament

Big, soft, and docile, Orpingtons can make very rewarding pets. They are easy to tame and very quick to trust their keeper.

Egg

Primarily the Orpington was developed with a brown egg in mind, but a number of strains is now tending toward tinted eggs, particularly when compared with true brown-eggers like the Welsummer (see page 84) and Marans (see page 72). Orpington hens can be very productive if managed correctly, although they also have a propensity to broodiness.

Plymouth Rock

Origins
The first Plymouth Rocks were bred in the US in the 1870s and were named after the Massachusetts town. The barred plumage variety was the first to be created, and the breed was among the first admitted into the American standard. It is believed the breed was developed from many others, including the Java, Brahma, Minorca (see page 74), Dominique, and Cochin.

Features
The Plymouth Rock is a fairly large breed of chicken, and as such it can easily be bred with a dual-purpose function in mind. It has both good laying and table qualities, and is raised in a variety of colors beyond the classic barred, black, white, and buff. Silver-penciled, partridge, Columbian, and blue are also seen.

Hen
The Plymouth Rock hen should weigh around 7.5 lb. (3.4 kg). She should have a deep, full-looking abdomen as if to show off her laying prowess. The breast should be broad and well rounded, in contrast with the small, compact comb and wattles.

The Plymouth Rock appeared on the breed map quite early on and is frequently linked, in its barred form, with the creation of a wide range of other breeds. It is a vigorous bird that lays a good number of eggs, but despite this, its popularity has ebbed and flowed outside of its home country since its initial fortunes.

Cock

The Plymouth Rock cock should be a large bird, weighing a hearty 9.5 lb. (4.2 kg). He has a bold, upright appearance that exhibits his solid, well-curved frame. His head is topped with a medium single comb, and he has medium wattles and large, lively, richly colored eyes.

Care

As is the case with many of the laying breeds, the Plymouth Rock will thrive in a free-range environment but can also cope perfectly well within a fixed run. The birds are capable of short flight, although it is usually the lighter hens rather than the much heavier cocks that are seen perched on a fence or roof. A robust and healthy breed, it copes with most weather conditions and extremes.

Temperament

Placid and friendly, the Plymouth Rock is easily tamed and makes a pleasant addition to a mixed flock. Among themselves, the cocks are tolerant, and given sufficient hens and space they will get along without any aggression.

Egg

The Plymouth Rock lays tinted to cream-colored eggs in good numbers, putting the breed among the top in laying trials. The hens can become broody and make good mothers if allowed to rear a clutch.

Rhode Island Red

Origins
The Rhode Island Red is an American breed and takes its name from the state in which it was formally created. Originally developed as a breed in the late 19th century from a wide selection of Asiatic fowl, it eventually had its Standard of Perfection documented and accepted by the American Poultry Association in 1904.

Features
Also known as the "Rhodie," the Rhode Island Red is well known for its strong mahogany-red plumage. The breed was originally developed with a rose comb but is seen today with single and rose combs, depending on the variety; both types are acceptable in the Standard of Perfection. It is a reliable layer that also has good table qualities. There is also a lesser known, less populous white-plumaged variety known as the Rhode Island White.

Hen
The Rhode Island Red hen weighs in at 6.5 lb. (2.9 kg) and is a solid, attractive-looking bird with a broad back that runs smoothly into the tail. The latter is carried low, giving the bird an apparently longer appearance.

The Rhode Island Red is one of the most well-known breeds of chicken in the world. It is also one of the most influential chicken breeds within poultry farming, having made a significant contribution toward many hybrid types used within the commercial sector.

Cock

The Rhode Island Red cock exhibits the underlying table qualities of the breed by weighing in at 8.5 lb. (3.9 kg). With an alert, active, and well-balanced appearance, the males clearly display the hardiness of the breed.

Temperament

Rhode Island Reds tend to be calm around their keepers. Unlike a number of the other laying breeds, they also can become quite trusting, with little desire to take to the wing.

Care

A hardy breed that is well adapted to extremes of climate, the Rhode Island Red works well within an enclosed run but is equally effective in a free-range setup as it has good foraging skills. Young birds grow fast and pullets come into lay quickly, with their first eggs being laid at around 20 weeks of age. Their combination of temperament, high egg production, and ease of care make the Rhodie an excellent breed for the beginner, although roosters can sometimes become aggressive.

Egg

The eggs of the Rhode Island Red are light brown in color and, as befitting this influential laying breed, the hens are capable of producing up to 300 eggs in a year. Their egg production will remain high for a number of years, making them an excellent investment for both large- and small-scale poultry keeping.

Welsummer

Origins

The Welsummer takes its name from the village of its origin, Welsum, which sits on the River IJssel in the Netherlands. The exact chicken breeds that make up its parentage are not fully known, but there is evidence of Cochin, Wyandotte (see page 86), and Leghorn (see page 70), among others, in its makeup. Standardized by the Dutch in the mid-1920s, it was exported to other countries from the 1930s onward but not added to the American Standard of Perfection until 1991.

Features

Present mainly in red partridge coloration but also seen in silver duckwing and gold duckwing, Welsummer sits at the heavier end of the light breed scale. A well-balanced, solid stance, married with a long, broad back, full breast, and moderately sized wings, provide the makings of a good example of a country or farmyard fowl.

Hen

The Welsummer hen weighs 6 lb. (2.7 kg). A medium-sized, neat single comb adorns her head, and equally well-ordered wattles hang from her face. She exhibits her laying prowess with a full, deep rear.

The Welsummer is renowned for the speckled, dark-colored eggs it lays, along with the very typical rich, warm "jungle fowl" plumage it exhibits. Well known as a prolific egg-layer, the breed includes some strains that are also capable of supplying a reasonable amount of meat for the table.

Egg

The eggs of the Welsummer can vary from a light brown through to a rich terra-cotta color that is prized by breeders. Some eggs display darker brown speckling on the surface.

Care

Welsummers are best suited to a free-range environment as they have the excellent foraging skills befitting of a country fowl. They will cope with the confinement of a fixed run, but suitable fencing or roofing will be required as they are quite capable of short flight. They are good layers, but they are known to be less productive during the winter months than some of the other breeds in the category.

Temperament

An upright, active, alert breed, the Welsummer is also admired for its friendly, gentle character.

Cock

The Welsummer cock should weigh 7 lb. (3.2 kg) or more. His plumage is markedly different from that of the hen, and although his headgear is larger, it remains in balance with the overall look of the bird.

Wyandotte

Features

The profile is characteristic. The body is short and deep but also well rounded, accentuated by the short back, broad saddle, and full breast. The Wyandotte is an alert and active-looking bird with mildly abundant, but not overly fluffy, plumage. A wide range of color is seen in the breed, including white, buff, black, blue, partridge, Columbian, silver penciled, silver laced, gold laced, and blue laced red.

Origins

The precise makeup of the Wyandotte is difficult to ascertain. It is an American breed that was first reported in 1860s, but little is known about the breeds that were involved in its creation. The first variety to be seen was the silver laced, and this sits well with the widely held belief that a breeder had initially set out to create a Sebright-plumaged Cochin. The breed was standardized in 1883 and additional color varieties emerged over the following decades, further serving to diversify the parentage of the breed.

Hen

The Wyandotte hen when fully grown should change to weigh around 6.5 lb. (2.9 kg). She has a graceful appearance and complements the alert presence of the male within the flock. This is heightened by the soft-looking plumage and slightly shorter stance. Hens are, however, prone to going broody, so if a high level of egg production is required, then appropriate management techniques are needed *(see pages 152–5).*

The Wyandotte is a very popular breed, notable now particularly for its wide range of available colors. Its large size and rounded shape, coupled with its full and profuse feathering, make it immediately recognizable. Some say it was named after a tribe of Native Americans, while others say the name came from a boat that belonged to the father of one of the Wyandotte's first breeders, Fred Houdlette of Boston, Massachusetts.

Cock

The Wyandotte cock is a proud-looking bird with a full breast lifted high, a neat rose comb upon the head, and medium-length wattles hanging from a clean, smooth-textured face. When fully grown, he should weigh around 8.5 lb. (3.8 kg).

Temperament

Wyandottes are trusting birds that will soon become friendly toward their keeper. This makes them an excellent breed for the beginner or for someone looking to keep chickens as pets.

Care

As a strong and vigorous breed, the Wyandotte copes well in all climates and most husbandry environments. The birds are not known to be keen flyers, and although they capable of taking to the wing, they can be contained by a low fence if kept in a closed run. It is worth noting that the housing provided for these birds needs to be in proportion to their slightly larger frame.

Egg

The Wyandotte lays eggs that range from tinted to light brown. While primarily a laying breed, it is not as productive as some due to the heavier breeding focus on plumage variety. That said, certain strains are capable of producing more than 200 eggs a year.

3

KEEPING CHICKENS FOR EGGS

CHAPTER FIVE

Getting Started

Keeping chickens can be tremendously fulfilling in so many ways. No matter what your motivations are, it is extremely hard to deny the anticipation of, and excitement in, finding the very first egg, laid by your very first flock, sitting in the nestbox, fresh and still warm. It's something that most small-scale chicken-keepers never fail to appreciate. They tend and care for their flocks of feathered farmyard assistants each day, and each day they are rewarded with eggs. It's a unique relationship and one that is almost symbiotic: The keeper provides feed, shelter, and protection; in return, the hen provides a packaged parcel of protein and goodness.

Why Chickens?

This is a good question. You might have just returned from a poultry show and seen a huge range of chickens exhibited. Alternatively, you may have dropped in at one the many small poultry businesses or farm-supply stores that can be found in most regions, or perhaps one of your friends or neighbors has been talking to you about the huge enjoyment and delicious eggs they are getting from their new flock of feathered friends. Either way, you seem to have been bitten by the bug and are excited about the prospect of keeping some chickens of your own.

This is understandable, since chicken-keeping can be a very fulfilling pastime. Not only are chickens extremely engaging creatures to have around, but there are different breeds to meet a wide variety of needs and, better still, to fit within almost any sized property.

Keeping chickens is an extremely scalable hobby that does not need to cost the earth or be totally time-consuming. A handful of hens in a homebuilt coop and run with a decent amount of space and a caring keeper will produce a plentiful supply of nutritious eggs for the kitchen of an average household. You may, in fact, be able to manage the flock and its expenses so that the venture works out as a cost saver compared with the price of purchasing eggs from the supermarket. That said, you will need to be prepared for hard work, as owning livestock brings with it responsibilities and the need for an understanding of husbandry beyond that required to keep a cat or dog.

So, before you start researching and looking at pictures of the myriad chicken breeds out there, stop, take stock, and ask yourself a few simple but vitally important questions. The first of these is: Why do I want to keep chickens?

This is an important question to ask at the outset to ensure that you get the most from the experience. You may be looking to have fresh eggs or you may want a supply of home-reared meat, or both. It might be that your family wants a pet and chooses to have one that is productive. Alternatively, you may want to get into the hobby of showing to become part of the local animal-breeding and exhibition scene. Or perhaps keeping chickens is a lifestyle choice to complement the vegetable patch or allotment you already manage.

The reasons are many, so take your time. Understanding exactly why you want to keep chickens will help get your new hobby off to a successful start.

First Things First

Having thought through why you might like to keep some chickens of your own, you are still probably eager to press on with deciding which breed you are going to keep. First, however, there is another vital question you need to ask yourself: "How am I going to keep my chickens?"

This might seem like a silly question—surely it's simply a case of providing accommodation, food, water, and shelter; opening and closing the coop door; and collecting baskets of fresh eggs every day? In many respects it can be as straightforward as this, but only if you get the correct breed and setup for your environment and lifestyle. Otherwise, it can turn into quite a trial, which in turn will significantly tarnish the experience for you and quite possibly impact the levels of husbandry for the animals that fall within your responsibility.

If there are chicken-keepers in your local area, make contact with them and ask to visit. While there are plenty of books, magazines, websites, and online forums packed with information for the novice, firsthand advice given face to face is invaluable. In fact, if possible, visit as many places as you can and ask plenty of questions. One thing that chicken-keepers seem to have in abundance is the will and desire to talk chickens to just about anyone, particularly prospective chicken-keepers.

If you can't find anyone locally, then go online and find the contact details for your local poultry club. The members may not meet very frequently, but there will be a show in your vicinity at some point that you can attend. Not only will this provide firsthand experience of the chicken exhibition scene, but it will serve to expand your network of poultry contacts and allow you to pick the brains of a range of chicken-keepers. Being able to call or email expert poultry-keepers is invaluable for the novice, becuse although chickens may be relatively easy to keep, they do have their nuances, and these aren't always covered in the available literature.

Another consideration that can be overlooked is whether you can afford the upkeep of a flock of chickens. The financial aspects of chicken-keeping are covered on pages 97–8, but before you make the commitment of becoming a chicken-keeper, bear in mind that no matter what breed you elect to keep or what methods you deploy in their upkeep, there will be an associated maintenance cost, even if that is countered by income from their produce at a later date.

Time is another essential investment. The amount of effort required to keep different breeds doesn't vary significantly, but the amount of time spent looking after them can be affected markedly by the numbers of birds you have in your flock, the type of housing you select, and the husbandry methods you elect to use.

You will no doubt now be starting to formulate some level of the expectations associated with keeping chickens. These include your expectations from the birds and, to a similar extent, what the birds can expect from you. Having gone through the "how" of keeping chickens, you may then need to revisit the "why" if the answer to one of these means that the other isn't possible.

The process can be iterative, but by taking the time to ask yourself these simple questions of why and how, and then taking a closer look at the reality of what is required from you and your environment, and most importantly what you can offer your chickens, your vision will become more crystalized and you will be able to narrow down the breeds you might be able to keep.

SPACE

You need only look at the range of commercial chicken farming practices throughout the world to realize that chickens are remarkably resilient. At one end of the spectrum, they are packed into cages, whereas at the other end, they are free to roam in woodlands and pastures. Yet despite this apparent robustness, they are quite susceptible to stress, with overcrowding being one of the main triggers. It is essential therefore that you assess the amount of space you have available for your new flock in order to establish the optimum number of individuals within that flock.

Also bear in mind that different breeds have different needs. This isn't just true of the feeds

Below
When it comes to husbandry, different breeds can have very different needs, and this can influence the breed options.

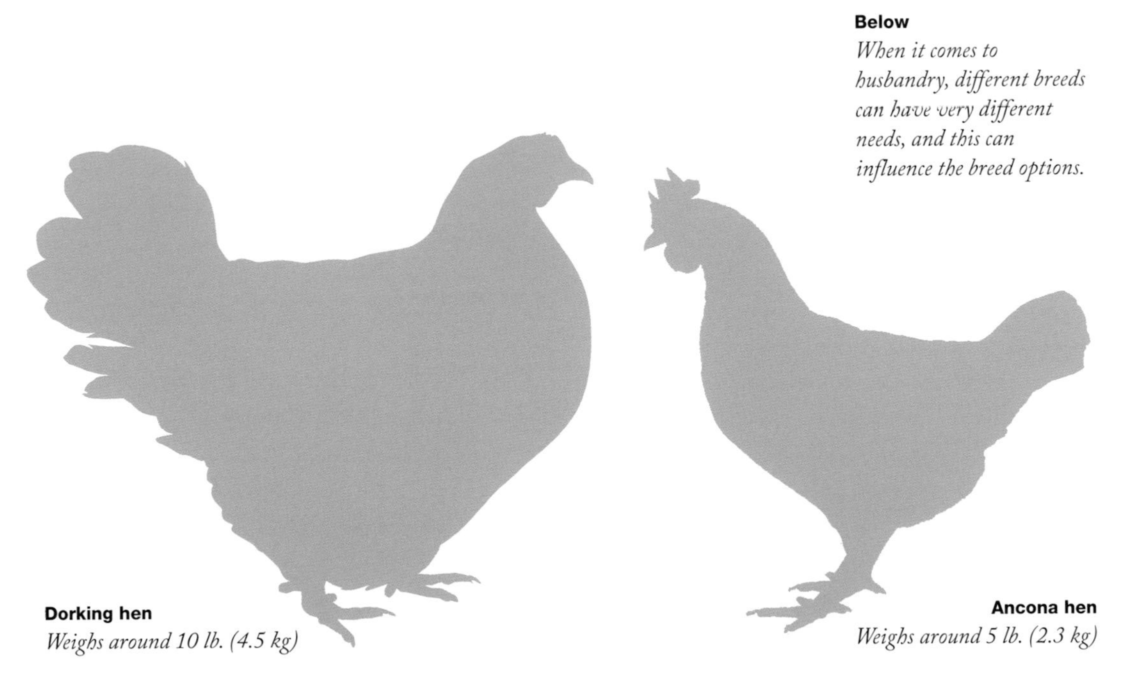

Dorking hen
Weighs around 10 lb. (4.5 kg)

Ancona hen
Weighs around 5 lb. (2.3 kg)

you use, the supplements you provide, or the type of the housing your birds live in, but also of the outdoor space you provide them. This might seem obvious when you stand a Brahma next to a Serama: the size difference is huge (possibly as large as you could get between two breeds of chicken these days), and as such it would seem sensible to conclude the former would need far more ranging space than the latter. But what if you stood a Dorking next to an Ancona? Again there is a visible size difference, but to conclude the Dorking needs more space than the Ancona would be incorrect. The former is a meat breed and generally a large and docile chicken that, while quite capable of ranging, cannot hold a candle to the frenetic energy that the light layer breed of Ancona applies to its everyday foraging. In this example, the Dorking would be quite content in a relatively small enclosure; the Ancona, however, would soon start to stress in the same space, being eager to explore further.

So what is the ideal amount of space you need to provide a flock? It's difficult to define precisely, but as a benchmark it is worth acquainting yourself with the local legislation regarding stocking densities for commercial flocks and then applying these to the proposed setup for your flock.

That said, space isn't everything. This might sound like a bit of a contradiction, but too frequently people equate more space with better, or higher levels of, welfare. This is probably due, at least in part, to the perpetuation in the popular media of the idea that cage farming in chickens equates to low welfare and free-range or organic farming to high welfare. This is misleading. Welfare is a measure of well-being, happiness, and health. Animals in high-welfare conditions are therefore happy, healthy, and well cared for; by the same measure, those kept in low-welfare conditions are distressed, sick, and unkempt.

For example, consider these two setups: first, a small fixed pen containing a trio of birds that is kept indoors under artificial light; second, a free-ranging flock of birds that roam for miles and love to roost outdoors. At face value, the latter seems to be the better welfare setup, but closer inspection shows the penned birds to be exhibiting all the correct behaviors expected from content and healthy animals, whereas the free-rangers are underfed and lice-infested birds whose love of roosting outdoors is due to the fact that their house is crawling with parasites and piled high with dung.

While the legislation at a commercial level will provide a guide, the backyard or small-scale keeper has a different set of challenges with a different set of advantages that enable husbandry to be adapted without damaging welfare, and in some cases directly improving it. There are three main types of setup you will encounter when looking into the care and husbandry of chickens on a non-commercial scale: free-range, fixed run, and movable run. It is possible to hybridize these systems; for example, keeping the chickens primarily in a fixed run but allowing them to be free-range at certain points of the day. It is also possible to swap and change during the year and as the seasons change—for example, free-range for the summer, and a fixed covered run during the wet or winter months. Each setup has its advantages and disadvantages, but if you have the time and space, it is perfectly possible to reap the best from all three. The table opposite gives a brief description of each setup, along with its associated advantages and disadvantages.

Chicken Husbandry Setups

Description	Advantages	Disadvantages
Free-range In the United States, there is no maximum stocking density, just the requirement that the birds have access to an outside area. Under European Union law the maximum stocking density for a free-range setup is 2,500 birds per hectare, or one bird per 45ft² (4m²).	Plenty of room for the birds to flourish. Better foraging, as the birds are able to range further, thus keeping feed costs down. Lower maintenance costs on fencing, runs, gates, etc.	Birds will range wherever they want. Good boundary fencing will be needed around your whole property to stop the birds from straying onto neighboring land. Increased exposure to pests and predators. Stock may elect to lay their eggs away from the house, possibly impacting productivity if not every egg can be found.
Fixed run A permanent, immovable enclosure within which the birds remain at all times. This tends to be a permanent poultry pen.	The birds are contained within a specific area and are easier to manage and catch. The birds will do less damage to your property, as they can't forage widely. The birds are easier to protect against pests and predators. Performance and productivity can more easily be monitored.	A higher maintenance effort is required in the management of the ground litter. Feed/forage supplements are needed, adding direct costs to egg production. Behavioral problems can occur, such as boredom or bullying.
Movable run This is a combined house and run that can quickly and easily be moved to a new location to minimize ground wear and tear. Movable runs sometimes come in the form of double- or single-tiered structures with handles, or they are built with wheels on one end so they can be moved like a wheelbarrow (usually called "chicken tractors").	Fresh ground can be provided as frequently as required. As fresh forage can be provided, there is less requirement for feed supplements. The location of the coop and run can be very flexible—for example, on a lawn during the spring to scratch out moss or thatch, on flowerbeds to clear insect pests in summer, or on vegetable beds after harvest to clear and fertilize the soil.	As the system is movable, it is usually much smaller in terms of the space available to the birds, so stock density needs careful consideration. The unit has to be light enough so that it can be lifted with the additional weight of your stock.

TIME

Depending on the number of chickens you intend to keep (which will be defined by your answer to the initial question of why you want to keep chickens), you will generally need at least 30 minutes a day to manage the requirements of the flock. This effectively breaks down as follows: 15 minutes at the beginning of the day to let the birds out, check their condition, fill waterers and feeders, and collect any eggs; and 15 minutes at the other end of the day to check again for eggs, check the birds over, and lock them up. These time periods are dependent on the season, but unless you invest in an automatic pop-hole door opener, somebody must be present about an hour after dawn and at dusk, 365 days a year in all weathers.

On weekends, you will also need about an hour per coop to muck out and perform the weekly tasks. If you frequently spend weekends away or take annual vacations, think through what you will do with the chickens then. Do you have friends or neighbors who can commit to the needs of the flock, or are there hen-sitting or boarding services nearby you can use?

The season or the point in the chicken's year also affects time demands, but one certainty is that if you genuinely take to keeping a flock of chickens, the time they require in terms of management may well remain the same, but the time you spend with the flock will significantly increase. They are incredibly endearing animals to be around, and once you start to understand the flock structure, its dynamics, the individual personalities of the birds, and the relationships between them, they will soon start to replace television soap operas or the latest bestsellers as your daily dose of drama.

SECURITY

Security here isn't about padlocks on gates or deadbolts on chicken coops, although these are sensible precautions when keeping some breeds or standards of chickens, as poultry theft is not unheard of. Instead it refers to the general design elements of the space in which you intend to keep your chickens.

If your intention is to keep a free-range flock, then the boundary barriers or fencing surrounding your property need to be considered, particularly if you adjoin other properties, roads, or public areas. Certain breeds of chicken are perfectly capable of jumping or flapping over barriers higher than 6 feet (2 m), so suitably tall boundary fencing is required to keep them in your own yard and out of any neighboring property or off any highways. If you don't intend to custom build or redesign the existing fencing and it's not of a suitable height, then this will restrict the choice of breeds or type of setup you can choose.

If, however, you don't intend to hand all of your yard over to your flock, then you need fencing that will keep the chickens in the part of the yard given over to them and not in the vegetable plot or sunbathing on the patio.

The fencing also needs to be secure enough to keep out any potential threats or visitors. Predators, and even hedgehogs in the UK, and raccoons in the US, will attempt to get into the chicken run either for the chickens themselves, for their eggs, or even for their food. As the first and often last line of defense, the boundary of the enclosure and the way in which it is installed plays an important role in the husbandry of your flock.

Threats can also be present from above. Birds of prey, such as buzzards, hawks, and

Right
Good fencing should aim to keep your chickens in and predators out.

kites, can take young growing stock, and members of the crow family are extremely clever egg thieves, dropping into the run, entering the house through the pop-hole, and leaving with an egg in the beak. To secure against this, roof netting or clear plastic panels might be a requirement.

Finally, the house itself needs to be secure enough to keep the chickens safe as they roost. The materials should be strong enough to prevent any predators scratching, digging, or chewing their way in. The construction of the henhouse needs to be sound enough to prevent it from being blown over by the wind, and any doors, nestbox lids, or hinged roofs should be secured so as to prevent them from being lifted by noses or paws.

FINANCIAL OUTLAY

The prospect of "growing" your own eggs is an exciting one, and if done correctly it can, both directly and indirectly, contribute toward the household budget. It does, however, require some cash up front—even the most frugal of startup systems will require a reasonable investment.

There are also the ongoing running costs to consider, although it is possible to balance these against the productivity of the flock. Keeping records of cost and income will help you keep a handle on the economics of your flock and will also enable you to establish a suitable price point for the eggs the flock produces if you decide to sell them, for example, at a farmers' market.

Startup and Running Costs

Initial outlay	
Housing and fencing	$500
Cleaning equipment	$70
Dietary supplements	$50
Treatments	$50
Total	**$670**

Running costs	
Feed (7 oz./200 g per bird per day)	8¢ per bird per day
Shavings (1/8 straw bale per week per house)	2¢ per bird per day
Total	**10¢ per bird per day**

OTHER FAMILY MEMBERS

It might seem like an odd question, but does everyone in the household like the idea of keeping chickens? The flock is likely to be around for quite a while, meaning that even the most hands-off member of the family will encounter them and perhaps need to adjust his or her lifestyle (if only slightly).

And are there any other pets in the household? Cats tend not to bother a flock of chickens much unless they are particularly aggressive, and creatures such as rabbits or guinea pigs are more likely to get picked on by chickens rather than the other way around. Dogs can, however, be a problem, so if you own a dog you need to give careful consideration as to how it might react to a flock of chickens and how you are going to manage the situation. While it is perfectly possible for the two species to coexist quite peacefully, there are some dogs that will never settle around chickens, constantly worry them, and in some cases kill them. If at all possible, find another chicken-keeper nearby and ask if you can test out how your dog reacts to their chickens—safely and on a leash, of course!

OTHER REASONS TO PAUSE FOR THOUGHT

Keeping chickens and introducing them onto your property or into your backyard might not be on a par with releasing wolves into Yellowstone National Park, but it does require you to perform an assessment of your immediate environment, and it should be given some careful thought.

The first consideration is the law. Are you legally allowed to keep chickens (often simply referred to as livestock) on your land? Do your land deeds accommodate this use and does the local legislation enable it, or are there rules that prohibit it? It's always better to check directly rather than assume you have the right to do so.

In a large proportion of cases, there will be nothing legally binding to stop you from keeping chickens on your property. Assuming this is the case and you have confirmed it, the next thing to consider is the "unwritten law." This is the one that is forgotten quite frequently—the law of common courtesy, decency, and respect, in which you take into consideration your neighbors.

You might be lucky and have no nearby neighbors to worry about, but for a large majority of people, neighbors are a fact or feature of our lives. As such, there are scenarios (such as when erecting buildings and additions) when you need to get your neighbors' permission first, or common sense dictates that it's worth getting them on your side before you commit to your project. You can, of course, simply ignore your neighbors, and if you have a fractious relationship with them already, then that might be the route you take. However, if that is the case, don't expect them necessarily to ignore your new feathered lodgers.

One way to ensure that you mitigate the risk of offending your neighbors is to visit someone, particularly a friend if possible, who already has chickens in his or her yard and, while there, perform the "eyes, ears, and nose test." Most people, when faced with the prospect of a neighbor getting a flock of chickens, will primarily be concerned with how it will impact them, and are likely to ask three questions: "Will they be an eyesore?", "Will they be noisy during the day and in the early hours?" and "Will there be a smell?" When you visit your friend's flock, ask these questions, see what the responses are, and test them with your eyes, ears, and nose. That way you should be able to give a confident and educated response should you get a similar set of questions from your neighbors.

One final thing to contemplate, and it could well be something your neighbors raise too, is uninvited visitors. This could be the cat from down the road, the dog from over the fence in the park, or predators such as foxes or raccoons that frequent your plot without your knowledge. If you do get regular uninvited visitors, you will need to ensure that you protect your flock accordingly. Your flock may also attract rats, mice, and, surprisingly, humans. Even the best-kept chickens will attract vermin, but it's the strategies you deploy when you become aware of their presence that will define whether they are dealt with efficiently or allowed to become a larger problem. With regard to uninvited human visitors, a lot will depend on the types of chicken you keep. As with any prized pets or livestock, if their value is high enough, then someone will always be tempted to steal them.

Above
Some roosters can be quite noisy throughout the day as well as at first light.

CHAPTER SIX

Pitfalls & Potholes

It is always preferable to go into any venture or hobby with your eyes wide open and having carried out as much research as possible. People head into chicken-keeping for a variety of reasons; the table on the following pages lists 10 such reasons with 10 possible potholes and pitfalls that can, and frequently do, befall those who don't do their homework.

The motivations for keeping chickens will have a significant impact on the decisions made regarding the type of chickens you will own and the manner in which you will keep them. In fact, when you go through the whole thought process, you might find your intentions cannot be realized with the resources you have. As such, you might need a bit of a rethink, or you may need to adjust your expectations. There's usually a way around most obstacles but it's important to be mindful of any potential limitations up front and before committing. It is essential for both your sake and for that of the poultry that you carefully consider why you want to keep chickens. Get these decisions right and clear in your mind first, and your chicken-keeping experience will be both valuable and fulfilling.

Top Ten Reasons for Keeping Chickens

Purpose	Reason	Pitfalls and potholes
Fresh eggs	The fresh eggs from your own flock of chickens will be the best you have ever tasted and picking the right breed will ensure that you have a plentiful supply, which, coupled with correct feeding and care, should keep the household well supplied with a food whose origins you know.	You pick a breed for its looks and not for its egg productivity, and it ends up costing you more to feed the hens than they return in eggs.
Home-reared meat	Chickens are an accessible form of livestock, and, depending on where you live, you may be able to butcher your own chickens at home. Check your local laws carefully, though, especially if you live in a residential area. More and more people see home-grown chicken meat as a better option than supermarket offerings.	You become attached to the flock and find you are unable to have the chickens butchered, let alone eat the meat.
Eggs & meat	You like eggs and appreciate home-grown meat, so why not keep chickens for both purposes?	You read up on dual-purpose breeds and settle for the one you think will suit you best, but you source poor-quality stock from an unproven utility strain that provides neither eggs nor meat in any real quantity.

Top Ten Reasons for Keeping Chickens *continued*

Purpose	Reason	Pitfalls and potholes
Pets	Dogs, cats, goldfish, rabbits, and the like do tend to figure high up the pet charts, but chickens are increasingly making an appearance—and why not? They are intelligent, sociable, friendly, and best of all, productive.	You pick a breed for its looks and not for its temperament. You handle the birds on the day you put them into your chicken house, but because they are a flighty, nervous breed, you never get within touching distance of them again for the remainder of their lives.
Conservation	Organizations such as the Livestock Conservancy in the US or the Rare Breeds Survival Trust in the UK focus on the conservation of heritage livestock breeds native to the country. Chickens can be accommodated in many people's backyards, and the contribution that small-scale chicken-keepers can make toward the conservation of their country's native chicken breeds can be massive.	You don't check the rare breed watch list closely enough or are not sufficiently familiar with the profiles of the breeds and end up conserving a non-native breed that is prolific in its home country.
Breeding & rearing	As more and more people want to keep chickens, supply needs to meet the demand, especially as not everyone has the capacity or knowledge to breed and rear chickens. It can also be a very fulfilling process.	You are overrun with cockerels because you failed to consider the fact that 50 percent of the chicks you hatch will be male, and in the majority of cases there is no market for these surplus males and you can't even give them away.
Exhibiting	A significant chunk of chicken-keepers are exhibitors who enter their birds into competitions. Breed clubs exist for a large majority of the chicken breeds, and members attend shows across the country and throughout the year in what is as much a social event as it is a competitive encounter.	If it's prizes you want, then you'll need to pay a premium for show-quality stock. Producing your own exhibition stock takes dedication and a lifetime of trial and error; success rarely happens overnight.

Top Ten Reasons for Keeping Chickens *continued*

Purpose	Reason	Pitfalls and potholes
Lifestyle	Not everyone has the ability or the resources to be self-sufficient, but everyone can be self-supporting in some shape or form. Chickens can make a significant contribution toward a more self-supporting lifestyle, especially when combined with a well-managed vegetable plot.	Livestock is a commitment; it's not a vegetable plot that can be left for a week or two while you take an annual vacation. Careful consideration is needed before incorporating chickens into your wannabe-homesteader dreams.
Business	You want to keep chickens as a money-making venture or incorporate them as part of a wider business plan. As the popularity of the hobby increases, so do opportunities to benefit financially to some degree.	Farming livestock for profit, no matter what the scale, takes skill. A lack of planning and a lack of understanding of the market forces in play will undo many ventures within the first couple of years.
Pleasure	Keeping chickens just for sheer enjoyment can be reason enough for some folk. Watching a flock wander through the yard, foraging for food, dust-bathing, preening, or just enjoying the sunshine can be reward enough without the need to consider egg output.	It's not all sunshine and dust-bathing—there's winter, mud, smell, pests, predators… In fact, a long list of unpleasant aspects that could crop up for the unwary.

CHAPTER SEVEN

Choosing Your Henhouse

With the huge increase in poultry-keeping, there has been an equally massive increase in the poultry paraphernalia that can be purchased, and nowhere is this more obvious than in the range of poultry housing. There are hundreds, if not thousands, of designs to pick from, with each one claiming to be the ideal accommodation solution. However, it cannot be stressed enough just how important the purchase of the henhouse is—not only is it likely to be the largest investment of your new venture, but it will be critical in ensuring that your chicken-keeping experience is positive and productive for both you and the flock.

Practicality is Paramount

Research and planning are key words in the process of getting started with chickens, and both are essential when it comes to selecting the type of coop you need. Visit as many stores as you can to get a feel for the products. If you have a large poultry show in your area, it may have trade stands where you can compare various manufacturers' offerings. If you have a local poultry club, contact the secretary to see if any members would let you visit to take a look at the types of housing they are using. Most breeders will be perfectly happy to share their experiences and help you avoid the pitfalls, so any legwork you can put in prior to parting with your cash will contribute significantly toward making the right decision.

Below
Visit people who already have chickens, as first-hand experience will help when making the decision.

Chickens require function over form, and unless you intend to open to the public, then you should also favor practicality over prettiness. A secure, dry, and draft-free area is the primary objective when a bird is looking for a place to roost, and this is the absolute minimum that a coop should offer. Beware of advertisements that offer attractive-looking houses at attractive prices photographed in attractive yards with equally attractive families standing alongside baskets of eggs. This is not reality and quite frequently the housing being sold in this manner is mass produced and not fit for purpose. Such houses will probably be constructed of fast-grown timber whose wide grain will swell in wet or damp conditions, resulting in jammed doors and pop-holes. Come the heat of summer, the timber dries, shrinks, and can even crack, leaving perfect breeding places for parasites such as the red mite (see pages 170–1). Think instead in a no-frills, utilitarian manner and look beyond the bells and tassels to the functionality of the coop.

You can, of course, create your own chicken coop, either by adapting a shed or outbuilding or by starting from scratch using one of the many plans that can found in books and on the Internet. The first option can be very cost-effective, especially if you have a storage shed that is not being used. With a little bit of DIY skill, it is quite simple to convert a shed into a suitable abode for your flock, and of course it has the added advantage of usually being large enough for you to walk into and stand up in.

With a little more DIY skill, it is possible to build yourself a henhouse from scratch. This can be particularly useful if you need to tailor one to fit a precise space—rather than having to compromise by installing a ready-made house, you are free to create one exactly to your specifications.

Whether you intend to research and buy a ready-built house, opt to convert a shed or other building, or dive in and develop your own bespoke property, the one thing you must do is understand the basic essentials of a poultry coop. They are not complicated structures, but there are certain "must-have" and "should-have" requirements that apply to almost any chicken coop. Recognizing these features and being able to spot when they are absent will help you evaluate the chicken housing on the market in your area or ensure that when you are modifying or building your own coop that you are always looking at things from a chicken's perspective.

Warning

Buying secondhand coops can be a risk. If your budget is such that you can afford only a secondhand henhouse, be sure to give it a thorough clean before putting any stock in it. Disease and parasites can sit in the dander and dust for many months, representing a very real risk to your flock.

Must-Have Housing Features

CONSTRUCTION

It cannot be stressed enough that good-quality, solid construction of the chicken coop is absolutely essential. Cheap housing is available, but it's really not cost effective in the long run. It invariably doesn't last much longer than a couple of seasons and will fail at some point, usually when it's least convenient, either due to the weather or general wear and tear. If it doesn't look like it could survive a small child standing on it or you leaning against it, then it probably won't survive a strong gust of wind or the weight of a winter snowfall.

VENTILATION

It is absolutely essential that there is adequate ventilation in a henhouse in order to reduce the risk of health problems within the flock. Roosting birds produce a large amount of heat, and they also produce a significant amount of feces. Without suitable ventilation, these two elements can result in a build-up of ammonia and, potentially, water vapor. The water vapor can then condense and either run down the walls to the ammonia-producing mass on the floor or drip onto the chickens in the coop, damaging the condition of their feathers and potentially chilling them, which could prove fatal during the winter months.

It is important to understand the difference between drafts and ventilation. The latter allows fresh air to be drawn into the building as the warmer air leaves. This is usually facilitated by having air holes at different heights on adjacent walls or a gap at floor level with a vent in the apex of the roof. A single set of holes on one wall will not facilitate air flow, nor will holes on opposite walls at precisely the same height. These would simply create a draft, something that should be avoided in chicken coops because not only will it stress the birds, making them restless and potentially less productive, but it can also trigger respiratory problems.

PERCHES

While not every breed of chicken, or in fact every individual chicken, makes use of perch bars, most birds will more often than not seek out somewhere to settle when they go to roost. For this reason, the inclusion of properly positioned perch bars certainly makes for an easier life for both the chickens and their keeper.

Chickens, and in particular the egg-laying breeds, will invariably seek out the highest accessible point in a coop each evening when they go to roost. If the house has no perches, then ideally the nestboxes should be at floor level. If the nestbox is higher than floor level, then the chickens will most likely use it as their roost spot. By the same measure, if perch bars are poorly positioned so that they are lower than the nestboxes, the chickens will ignore them and choose instead to roost in the nestboxes.

Tip

Ventilation is essential throughout the year. Don't be tempted to seal the coop up, even in the coldest of weather, as this will serve no benefit to the birds inside.

> **Tip**
>
> Perches for large or heavy breeds of chicken ideally should not be higher than 12 inches (30 cm) above the floor level of the coop. This is because these chickens run the risk of a foot or leg injury when they jump down, which can become infected and in turn develop into a painful and potentially fatal condition known as bumblefoot (see page 168).

This causes a problem because chickens produce a significant amount of feces while roosting, so consequently they will foul the area in which they lay, resulting in dirty and contaminated eggs. If, on the other hand, the perch bars are higher than the nestboxes, the chickens will use them as intended, leaving the nestboxes unsoiled.

A good perch will be 2–3 inches (5–8 cm) thick for large fowl and 1–2 inches (3–5 cm) thick for bantams. This will allow a bird's feet to settle almost flat on the surface with a slight overhang of the toes to provide extra grip. It will also allow for a reasonable amount of its keel to sit flat on the surface. The perch bar should be made from rough-cut square timber with the corners rounded off to a smoother curve. This will enable the chicken to get a suitable grip on the perch. Any perch bars made from sanded dowel or pipes made of plastic or metal are too smooth and will potentially be uncomfortable for the perching chicken, as the birds will not be able to grip correctly. A chicken that is not comfortable, relaxed, or secure when roosting can become stressed, and a stressed chicken has reduced productivity, so it is important to get something as apparently simple as a perch bar correct.

On average, each bird will need 8 inches (20 cm) of perch space (obviously less for bantam breeds and more for heavy breeds) and, if more than one perch is installed, the perch bars need to be at least 18 inches (45 cm) apart to avoid overcrowding, enable sufficient air flow, and provide adequate space between each row of chickens.

NESTBOXES

As a rule of thumb, a henhouse should contain at least one nestbox for every three hens, although it's highly likely that you will find that your chickens favor one or two nestboxes, and that's where you will find all the eggs each day. This, however, is more of a behavioral trait in some chickens as opposed to them not requiring the additional laying space.

The nestboxes ideally should be accessible from a door separate to the main access doors. Not only does this make egg collecting quicker and more simple, but it also means you can access the nestboxes without disturbing any chickens that are in the coop at that point. It also reduces the likelihood of you trailing dirt into the house, thereby improving biosecurity.

The size of the nestboxes should be appropriate for the size of birds being kept, as laying chickens like to feel safe and sheltered. Obviously the larger breeds will need a larger nestbox to avoid unnecessary damage to feathering, as hens frequently perform nesting behaviors before settling to lay an egg.

If the henhouse you are thinking of buying claims to accommodate 10 birds yet has only one nestbox, be wary—it is distinctly possible that the designer and manufacturer have missed other essential aspects, too. The same caution should be exercised when the house builder produces accommodations suitable for three birds and then adds three nestboxes. Overengineering of this nature may appeal initially, but it is possible that other basic errors will be present in the design of the house.

The ideal home

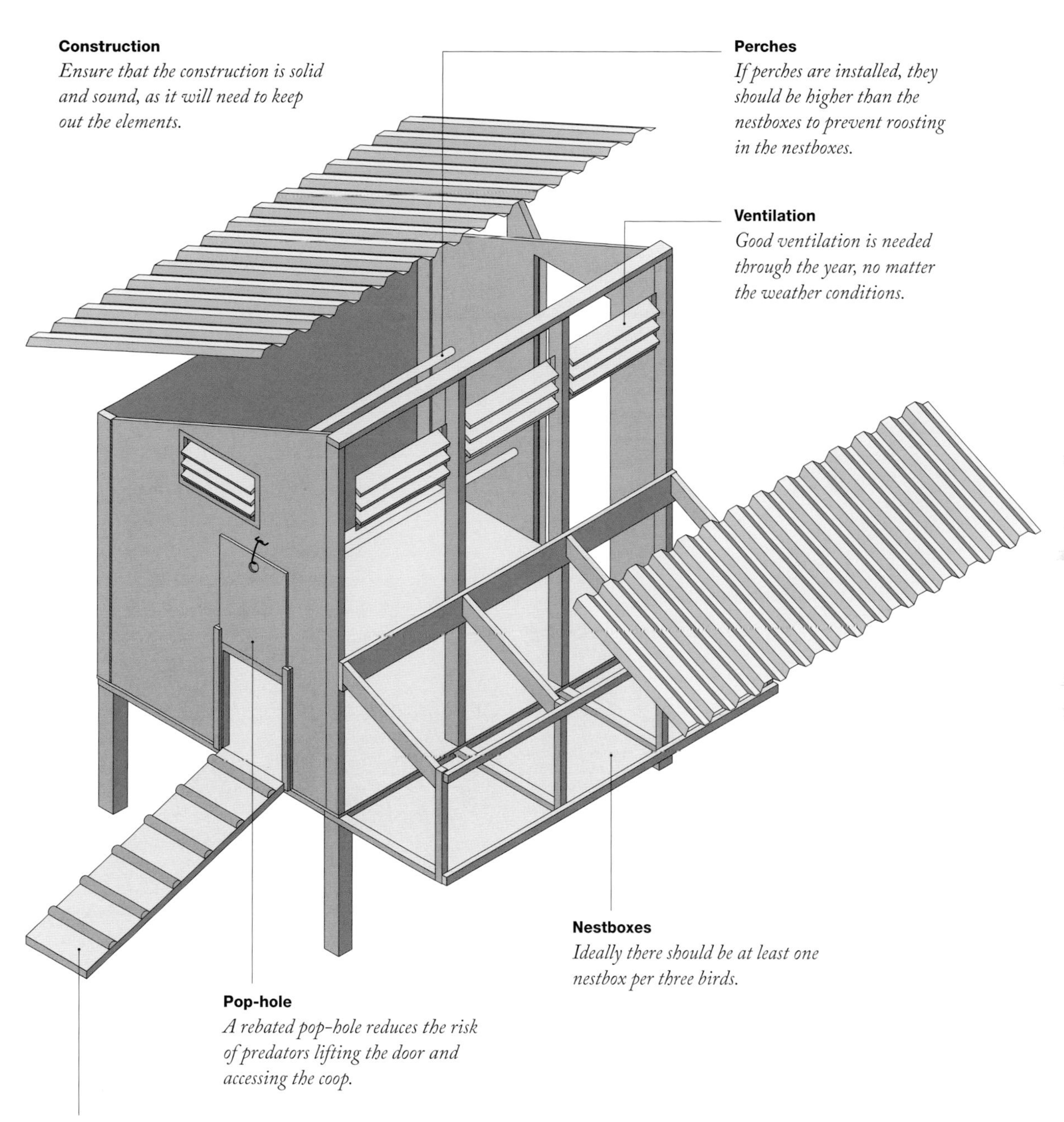

As mentioned in the foregoing section on perches, the nestboxes should be positioned at a lower height than the perches. If egg productivity is to be maximized and egg damage minimized, it's also important that the location of the boxes within the house is correct for the hen. Ideally, hens like dark nesting places, so ensuring the nestboxes are in the darkest recess of the coop will increase the likelihood of a relaxed, settled laying flock. If the nestboxes are directly opposite a window or opening, the hens could potentially become agitated. They may then lay while standing or stooped in the nestbox, resulting in the eggs breaking. Broken eggs not only equate to lost productivity, but they could make a mess of other eggs in the nestbox. Worse still, they may encourage "egg-eating" within the flock, a behavioral habit that is very difficult to break once established at an individual or flock level (see page 150).

POP-HOLES

The pop-hole on a chicken coop should be a simple mechanism and one that is easy to operate. Overengineered designs result in chains that jam, doors that swell and become wedged shut or open, and runners that fill with dirt and shavings, restricting the door's capacity to be securely closed. It is also important that the pop-hole can be secured shut with a bolt, or at least be of a rebated design such that the door's leading edge slots in between two other pieces of wood (or plastic). This will help prevent the door from being lifted by the paws or noses of any passing predators.

Left
To reduce the likelihood of nestboxes being roosted in overnight, they should be lower than perches.

ACCESS

Possibly one of the biggest oversights in coop design is access—not for the birds, but for you, the keeper. You will require access to the nestboxes in the house at least once a day to check for eggs, but, more importantly, you need access to the whole house in order to clean it out. You will need to be able to reach every corner of the house in order to ensure the cleaning is thorough and that any areas that could harbor mites or lice are properly attended to. Removable roof sections and wall or door panels are ideal in this respect and will enable you to tackle the cleaning job with both hands.

If the coop comes with an attached run, it is also essential that this is accessible so that you can reach every corner. Simply lifting the entire run off the ground will result in the chickens making a break for the nearest flowerbed and you wasting many hours trying to coax them back into the run. By the same measure, a run with insufficient access will equally result in wasted time as you try to salvage eggs laid outdoors that are just out of your reach or an empty waterer that's been kicked into a corner. Once again, if the house you are looking at has these design flaws, then it is distinctly possible that there is some other basic error elsewhere in the design.

SPACE

The final essential to consider is the capacity of the coop. Some housing may have the correct dimensions and length of perch within the coop, and the correct number of nestboxes for the volume of birds it claims it can accommodate, but always check the outdoor space if it comes with an attached run. In the US, there are no national regulations for poultry-keeping setups, but again, you should ensure that your chickens are kept in the best conditions possible. The European Union stocking density guideline for free-range chickens is one bird per 2 m x 2 m area (6.5 feet x 6.5 feet), barn-reared birds require 0.33 m x 0.33 m (1 foot by 1 foot) each, and enriched cage systems measure slightly over 0.27 m x 0.27 m (10.5 inches x 10.5 inches). If you live within this region, space restrictions in your yard may therefore mean that your birds will not be free-range. However, if your objective is to provide the best welfare for the chickens that are producing your household's eggs, then ensure that the henhouse and run allow sufficient room for each bird living there.

The chicken's nest

Bird nests come in a wide variety of forms: Some are built on platforms, others are hanging structures, some are cup-shaped and hidden away in the shrub, and others are in burrows or tunnels. Chickens create what are known as scrape nests. This type of nest is quite typical of ground-nesting birds and is perhaps the most minimalist of avian architecture. It serves one function—the incubation of the eggs—and because it is exposed to predatory threat, location takes over from design. That said, the nest is still a construction effort where the hen will find the right spot, lower herself into position, and shuffle her body and feet until a depression is made. She will then undertake a bit of sideways throwing and sideways pulling, during which she casts aside any unsuitable sharp material (stones or large sticks, for example) and pulls more desirable soft materials (feathers and leaves) towards and into the nest.

Should-Have Housing Features

As with many significant purchases, it's often nice to feel you've invested in something beyond the basic model and added some worthwhile optional extras. It is important, however, to remember that embellishments to a henhouse can often mean more surfaces to clean or, worse still, more areas in which mites can hide. That said, there are a few optional extras that, if included in the design, will enhance the coop and make things easier for you, the keeper.

REMOVABLE PERCHES

These can make cleaning much easier as they allow better access to the house, and they can also be replaced with fresh new perches periodically. Bear in mind that a chicken can spend a minimum of 10 hours out of 24 sitting on the perch bars, so it's important that they are maintained. In fact, having any removable components within the coop can be a real boon as they are much easier to clean and disinfect, and even simpler to replace.

RAISED HOUSING

A henhouse that is raised (or that can be raised) 6–8 inches (15–20 cm) off the ground is an optional extra worth considering. A raised coop has a number of advantages: first, the house is higher so you don't need to stoop as much to clean it out; second, it will deter vermin from taking up residence under the house; and, finally, the area of ground under the house provides a perfect shelter for the chickens from the rain and wind and will soon become a favorite dust-bathing spot.

INDOOR FEEDERS

Always aim to buy a coop that not only has enough space for the number of chickens you wish to keep but also to accommodate a feeder within it. Better still, choose a design with a feeder built into the wall that can be accessed and filled from the outside. Being able to keep feeders indoors not only stops the feed from becoming spoiled by the weather but also deters vermin and wild birds from helping themselves to the feed.

WOOD

Wooden housing offers greater flexibility and modification, being able not only to meet the specific needs of certain breeds but also to blend in with the local environment. Wood is a naturally breathable product, and a well-designed wooden house will not suffer from condensation. Wooden henhouses are also easier to repair should the need arise. On the downside, wooden coops take longer to clean out and dry than plastic houses, and they require more annual maintenance.

Housing Materials

While chicken housing is now constructed from a variety of materials, mass-produced coops still tend to use wood or plastic. If you are lucky enough to have outbuildings or intend to build a more substantial place of residence for your flock, then bricks and mortar are another option. However, the same design essentials still apply, and your needs and those of your chickens must be incorporated.

Most manufacturers of wooden housing tend to build a frame and then clad it with tongue-and-groove planks. As the name suggests, these are planks of wood that slot together, one side having a "tongue" that slots into the "groove" of the connecting piece to create a join. It's frequently listed as "T&G" and is very versatile to work with, enabling a wide variety of design and sizing options. By its very nature, however, T&G construction has gaps, which can be used as hiding places by mites and lice, so always make sure tight-fitting, good-quality T&G is used and avoid thin wood or flimsy construction .

Above
T&G is a popular building material for chicken coops, but it is not without its problems.

Tip

If you are intending to construct your own chicken coop from marine plywood, clad the house framework from the inside as opposed to the outside. This reduces the gaps in the wood even further and means you can seal the joints, thereby eliminating any potential hideaways for lice and mites.

> **Tip**
>
> Catches, clasps, hinges, and any other metalwork used in coop construction should be solid and ideally made from stainless steel. There is little point in having a top-quality henhouse with substandard metalwork that can rust or snap off.

The other wood option, and probably the best, is marine plywood. This comes in sheets and provides an extremely practical option to the poultry-keeper as it is strong, stable, and durable, and also gap-free. On the downside, it's not as attractive as T&G, and, as a result, its use within the mass-produced market is limited. For the home handyman, however, it is the better option, particularly if you are looking for function over form.

PLASTIC

Plastic sheeting is similar in many respects to plywood, but plastic housing in general has a limited number of design options. Either the sections need to slot together as opposed to being screwed or bolted into place or the coop is made in a defined injection mold. Add to this the fact that plastic is a relatively new coop construction material, and the result is that the size options on the market tend to be relatively limited.

Plastic house construction comes into its own versus wood when cleaning and performing annual maintenance, the latter is practically nonexistent when compared with a wooden house, as no weather treatment is needed. Cleaning a plastic house is also much easier than cleaning a wooden house; many of the designs can be cleaned out, pressure-washed, and dried in less than 30 minutes. This advantage should not be underestimated. In the height of summer, your flock may not require the house for the whole day, so cleaning and drying times won't be relevant, but on bad weather days and in winter you won't want the house out of action for too long or your birds having to roost in a soggy coop.

Many of the mainstream manufacturers of plastic henhouses are now endeavoring to use recycled plastics that are precision-cut, with any waste material returned to the recycling process. They literally don't cost the earth.

ROOFING MATERIALS

The material used on the coop roof is another important consideration. Many houses on the market use roofing felt, an old-fashioned weatherproofing solution for wooden roofing. However, while felt is perfectly suitable for this purpose when it comes to storage sheds or other outbuildings, it should be avoided when it comes to poultry housing for two main reasons. First, it's not as durable as would first seem the case. High winds can pick at a loose corner or small tear, resulting in the whole sheet being blown away. In addition, very hot weather can cause roofing felt to melt, and over time it can also become brittle and hard,

> **Tip**
>
> Make repairs to your coop as soon as you notice a problem. In the event that a hinge breaks, a latch snaps, a part the roof becomes detached, or any other relatively minor breakage occurs, be sure to fix it as soon as is practically possible. Leaving it can have disastrous consequences that compromise the safety or welfare of your flock—it's far better to do today what you were going to put off until tomorrow.

eventually cracking and causing leaks to appear. The second issue is that roofing felt frequently provides the perfect home for red mites, which will happily multiply in a place that is all but impossible to treat.

The frequently used alternative is corrugated roofing material, which comes in panels and tiles. These are very tough but lightweight products that can be nailed either onto a solid roof or directly to roof battens. They don't rot, and they are flexible, easy to work with, and they are weather resistant.

Above
If you mix and match the materials you use in constructing your henhouse, you can get the best out of all the options.

Basic Run Requirements

If you are not intending to allow your birds to roam free-range, then you need to provide them with a designated outdoor area or run. Even when using a free-range system, it is handy to have a compound where you can temporarily corral the birds, such as when you need to check the flock over. Obviously, different breeds potentially have specific needs, but there are some general requirements that should be built into the enclosure.

First off, be realistic about the amount of space you have. Chickens thrive on space; they are by nature foraging creatures and tirelessly scratch around for insects. If the space provided is limited or overpopulated, then it will soon be denuded of vegetation, insects, and tasty morsels. The ground will become littered with droppings that, come a heavy downpour of rain, will transform into a muddy mess, providing an ideal environment for disease. It can also cause behavioral problems as the chickens become bored and begin to peck at each other. This pecking evolves to picking and pulling of feathers and exposed flesh and can lead to full-blown cannibalism in extreme cases.

As the keeper, think about the location of the run from your own perspective. It might seem like a good idea to put the chickens well away from the house or main part of the yard when it's summertime, but what about in the depths of winter, when you have to trudge through snow, rain, and hail to reach them? The remote location might not be quite so appealing in those circumstances. Pick a location that is out of the prevailing wind and that gets some sunshine but equally has some shelter. As descendants of jungle fowl, chickens are happiest with some overhead cover, so if possible locate the run under established shrubs or small trees that will provide some canopy cover for them. Obviously, avoid damp areas—aside from the fact that chickens are not particularly keen on muddy conditions, a dry location will reduce the amount of mud and moisture being taken into the coop. And if space permits, divide the area into two or three smaller runs so that you can rotate the space occupied by the chickens, thereby giving each of the other areas a chance to recover.

FENCING

When considering the options for fencing, it's important to understand its purpose fully before embarking on any landscaping or purchases. Fencing needs to be functional and practical, but by the same measure you may prefer it to be complementary to the other landscaping within the yard. For example, if you have a more formal, orderly design to your yard, then running a mass of stock fencing

Left
Even with almost limitless space, chickens will invariably walk the boundary of a run.

across it isn't going to look the part. The other element to consider is whether the fencing will be a permanent feature, semipermanent, or temporary, as this will also influence the materials you use.

Brick or stone walls, if built high enough, work very well as a permanent boundary option, and with the variety of construction materials available, it is possible to create a structure that is in keeping with your property. Another permanent option is stock fencing. This comes in a variety of gauges or sizes, some with a tighter mesh closer to the ground to deter rabbits, a fringe benefit of which is that it also prevents large fowl chickens and many of the bantam breeds from getting through. Be aware, though, that topping stock fencing with a rail may help secure it, but it also means that you have offered a perch to any chicken that is capable of getting off the ground.

In terms of semipermanent options, movable panels made of woven hazel or willow can work very well. These provide an attractive way of partitioning an area of yard and are available in a number of sizes (height and length). The down stakes on the panel can usually be driven into the earth, and then cable ties or heavy-duty twine can be used to tie the structures together to give them a bit more strength. Due to their lattice-like construction, these wooden panels also offer a good level of resistance against strong winds and act as a windbreak to protect your flock. Should you want to relocate the fence, simply untie the panels and move them to their new position. On the downside, they can be quite expensive to purchase, and while they will last a number of seasons, they do not last indefinitely.

Another fencing option comprises wooden frames covered with chicken wire. These are certainly functional and a good way to separate birds temporarily or to divide paddocks into smaller breeding pens. They are not as attractive as woven wood panels, but they do last longer.

Temporary options include plastic windbreaks and electric fencing. The latter doesn't need to be electrified if it's being used as chicken-yard fencing as opposed to antipredator fencing, a factor that can keep the investment costs down. Both fence types are very handy as they can be rolled up for storing.

The type of fencing you choose depends on how secure and/or how permanent or flexible you need it to be. In most cases a 4-foot-high (1.2 m) fence will serve as a suitable barrier to most breeds of chicken, and if used in conjunction with wing-clipping for the more flighty breeds (see page 137), it should keep the birds contained. That said, make sure you do your research, because next to the chicken coop, fencing can be quite an investment in terms of money and effort.

Other Essential Items

In addition to housing and fencing, you will need to buy some equipment before you get your birds to aid you in managing your new flock. There are a lot of items of poultry paraphernalia on the market, all of which aim to make the job of keeping chickens easier, but before you commit too much of your budget to optional extras, make sure you have the following essentials.

WATERERS

Plastic waterers are cheap and easy to clean, and they don't corrode if you choose to add apple cider vinegar to the water (see page 164). Metal waterers are more durable, but they can be expensive. Always select a design that is stable so that the chickens can't knock it over easily (and consequently find themselves without water) and preferably raised off the ground. Chickens naturally scratch, so a low-level waterer can easily have dirt and feces kicked into it. The size of the waterer will depend upon the size of the flock and also the time of year, as chickens vary the quantity they drink according to the season. Aim to provide at least 2 pints (1 liter) of water per bird per day and at least one waterer for every five birds to reduce competition.

FEEDERS

There are many different designs and materials available on the market, all with advantages and disadvantages. The best feeders are those that keep the feed in an area where the chickens are unable to scratch it out onto the ground, either by having a mesh or grid over the feed, by raising the feeder so that it is too high for the chickens' feet to reach, or by using a treadle feeder. The latter keeps the feed covered until the chicken steps on a plate to enable access to the feed, and has the added advantage of keeping out wild birds and vermin. One other important consideration is to make sure the feeder is large enough to carry sufficient feed for your entire flock for the day, or preferably longer.

COMPOST HEAP

Chickens produce a large amount of manure, which, when combined with the litter on the floor and in the nestbox of the henhouse, can accumulate into quantities beyond that which your household refuse bin can cope with—not that it should be simply thrown away. Chicken manure is an excellent accelerator for the compost heap, and when combined with the litter and other compostables, it can create an excellent booster for your garden soil.

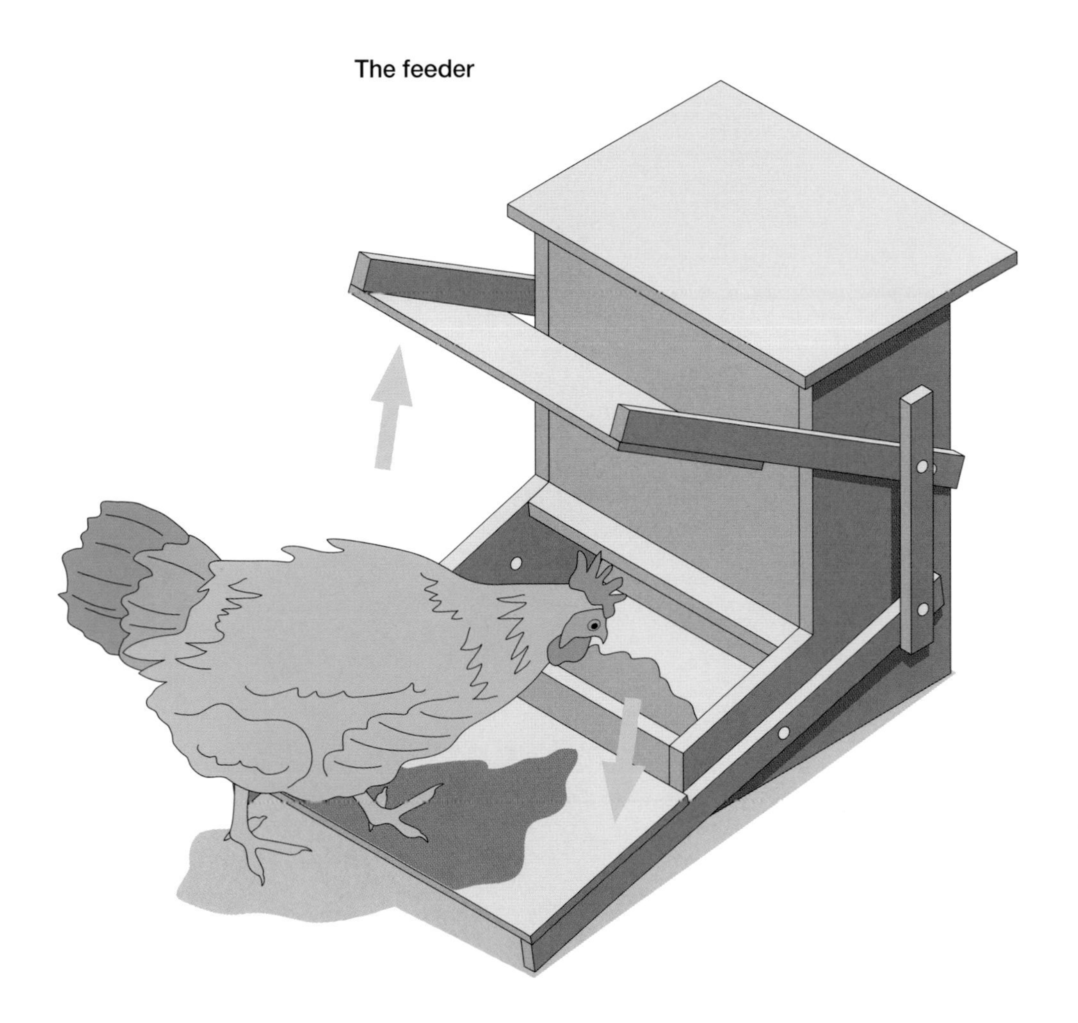

CLEANING EQUIPMENT

For reasons of hygiene, buy cleaning tools used solely for chicken-keeping. Depending on the size of your flock or the size of the coop you are using, the equipment you need can be as simple as a dustpan and brush, a wallpaper scraper, and some mucking-out tools.

Above
Treadle feeders are a good choice as they can reduce feed waste and loss to wild birds or vermin.

EGG BOXES

It's perfectly acceptable to use old egg boxes for storing the eggs you are keeping for your kitchen, but if you intend to sell eggs at the gate or at a market, you will need new boxes.

GRIT

Chickens need a supply of grit to help them digest their food (see pages 160–1). Even if they are free-ranging and finding their own, it is important to provide them with a tray of grit.

CHAPTER EIGHT

Your Egg-Laying Flock

Chicken breeds loosely fall into six categories or classifications in terms of their function or purpose: layer, table, dual-purpose, true bantam, game, and ornamental breeds. There is some blurring of lines between the different breed types and there are even breeds that have made a transition from one type to another at some point in history, or could arguably be placed in more than one group.

There is also another widely recognized division between pure (or heritage) breeds and hybrid chickens (see table below). Pure breeds tend to have been developed through many decades of careful breeding. The definition varies according to different countries and different organizations, but as a general rule of thumb such breeds have been recognized since before the mid-20th century (pre-1950).

A hybrid is an intentional crossing of different breeds of chicken with a specific end game in mind of either fast-growing table birds or more efficient and cost-effective egg-layer birds. Hybrids have generally been developed since the 1950s, which was a key time period in the development of many of the commercial strains we now encounter on large-scale poultry farms.

Chicken Classification—A Quick Guide

Description	Pros	Cons
Hybrid Commercial strain developed after 1950	Inexpensive to buy Predictable nature Heavy egg-layers Usually vaccinated Fast-maturing	Short lifespan Laying difficulties Limited breeding use Function-specific (eggs or meat)
Pure breed Heritage/old breed that originated before 1950, or a non-commercially developed breed	Long-lived Variable egg color Variety of plumage types Variety of sizes Variety of natures Some function-independent breeds	Can be expensive Difficult to source Egg output is dependent on breed Not usually vaccinated Slow-maturing

Pure Breeds & Their Specialties

Type of breed	Characteristics	Care & management
Layer breeds	Excellent layers, producing large numbers of eggs Light birds, often the smaller of the large fowl breeds Fast-maturing Excellent foragers Spritely character; not too easily tamed Non-sitters, so don't go broody often	Capable of short flight, so suitably high fencing or a roofed-in run is required Clean-legged, capable of coping with wet conditions
Meat breeds	Heavy, large birds, with males often exceeding 10 lb. (4.5 kg) in weight Not very active Non-flighty Placid nature/non-aggressive Poor layers compared to layer breeds Often go broody and make excellent mothers	Low fencing is sufficient Not far-ranging; happy in a smaller space Big appetite, so care is needed to prevent them from getting fat Accommodations need to be appropriately proportioned Low perches needed to avoid leg or foot problems Large nestboxes required
Dual-purpose breeds	Reasonable layers with reasonable market weight Durable breeds that cope well with poor weather Can be tamed with patience Generally placid and tolerant of other breeds Some go broody, others are non-sitters	Care and management vary with each breed because different breeds have different characteristics and hence different requirements

Pure Breeds & Their Specialties

Type of breed	Characteristics	Care & management
True bantams	These breeds have no large fowl counterparts and occur only in a bantam form Tend to be ornamental in appearance "Showy" appearance Not known for egg-laying Broody, so make excellent mothers Proud cockerels, can suffer "little man syndrome"	Can be kept in a small enclosure Great for small properties Very good for young children Housing needs to be sized accordingly Roof may be required as some breeds are flighty Need a dry run, as their short stature can result in dirty feathers in wet conditions
Game breeds	Self-confident and assertive Very intolerant of other breeds or birds outside of their flock Non-flighty Can become extremely trusting toward the keeper Hard-feathered, with posture more important than plumage Strong and dominant Very poor layers	Extremely tough and durable Require secure housing and must be kept separate from other birds Need suspended waterers and feeders due to their upright posture Some breeds are monogamous
Ornamental breeds	Grouped because of their accentuated features Generally rare and specialty show birds Carry attributes of the other groups Qualities are very breed-specific Usually very striking-looking breeds	Very variable husbandry requirements Care and housing must be tailored according to breed Can be a challenge to show successfully

Chickens for Eggs

Chicken breeds that fall within the layer category, be they hybrid or heritage, are capable of producing large numbers of eggs. They are usually smaller in size than those that are bred for meat or those that sit within the dual-purpose category. The reason for this is simple: they put their energies into laying eggs, not into putting on body mass. They also tend to come into lay sooner than the other types of chicken, the intent being to have them reaching maturity by 20 weeks of age and to lay as many eggs as possible within the first few years.

Layers are often excellent foragers, scratching and digging industriously for additional tidbits, a trait that contributes to their efficiency by minimizing the feed costs for maximum output.

Their nature is often sprightly, though shy, and some breeds can be very wary of their keepers, who need to show a patient and relaxed approach if they are to tame the birds successfully.

Many of these breeds also fall into the "non-sitting" category, meaning that they rarely go broody. The trait of broodiness is not favored in laying breeds, and over the centuries it has been selected out through careful breeding. If you intend to raise laying breeds from fertile eggs, alternative hatching methods may be needed, such as an artificial incubator (see pages 30-34) or a willing broody hen of a different breed (see page 154–5).

The key quality of a layer breed is the volume of eggs it can produce. For purebread layers, this is often viewed as the ability to lay more than 200 eggs a year (ideally more than 280 for the best individuals in a flock). For commercial hybrid strains, this figure exceeds 300 eggs in the first year of production (but achieving this needs careful management of the environment and can have health consequences for the birds themselves).

Historically speaking, up until fairly recent times, almost all domesticated chickens were viewed as egg-layers. The primary purpose of the chicken was to lay eggs, and lots of them. This took priority over the provision of meat for the table, and only when the bird ceased being productive would it be culled and used for food. Many countries and regions carefully developed their own specific purebred egg layers to suit their local environment. As a consequence, there are a wide range of heritage or pure breeds with an equally wide-ranging history. Some, such as the Lakenvelder (see page 66), date back to the early 18th century, while others, such as the New Hampshire (see page 76), emerged as a breed only in the early 20th century.

Left
Hybrid hens are frequently used as the main egg layers in a flock.

Tip

It is important to note that the egg output of a particular hen can also depend on its bloodline. Many of the breeds, or colors within a breed, that were once excellent layers have changed over the years, having been developed for exhibition—examples include the Leghorns and Orpingtons. Although individuals and bloodlines within these breeds may appear the same, they might have lost their high egg production, so check with the seller before you purchase any stock.

In general, layer breeds tend to be lighter in the body, more agile, and, in some cases, quite capable of short flight. As such, high fences or roofed-in runs need to be considered to stop them from straying too far. They are predominantly clean-legged, so will cope better in wetter conditions than those with foot feathering.

The breeds are designed to lay daily for as long as the daylight hours allow, so it is good breeding practice to weed out the poorly performing individuals within a flock if you intend to breed replacements. Birds that have good feathering toward the end of the season or those still laying eggs with heavy pigmentation will be infrequent layers and should not be bred from. In contrast, a bird that has laid well during the season will have shabbier feathering and a decreased ability to color her eggs at the end of the season. Her productivity will make her a good hen to breed from if egg production is the sole objective.

Heritage Breeds & Hybrids

The Victorian craze for fancy fowl in the mid-19th century significantly changed the Western perspective on chickens. In fact, it undeniably put poultry in a new light within modern culture. No longer were they simply a "farmyard forager"—they now had a different and wider appeal, which was reflected in "hen fever" cutting across the classes of the Western Hemisphere.

The influx of new breeds not only delivered aesthetic appeal and price tags beyond anything seen before, but also egg-laying and meat capabilities that raised economic eyebrows both at homesteads and, later, at a commercial level. It was both of these practical functions that would become more pivotal over the next 100 years.

The austere times that were brought about by the two world wars put meat among many other things in short supply and it was the chicken fanciers of the time who, unknown to them, held the answer. They had already developed certain breeds and strains that had the potential to provide plenty of eggs and meat, and the fact that such development was occurring in a non-commercial manner meant the path of the humble chicken (and chicken breeder) now split into distinct directions: one as an exhibition hobby, tipping its hat to the late 19th century; and one as a potentially mass-produced food keenly needed by the world's population.

No more so was this better defined than in the late 1940s, when a collaboration between farmers, breeders, and suppliers, backed by cash prizes from the A&P supermarket chain, launched the Chicken of Tomorrow contest across the US. The concept behind the contest was to challenge the huge numbers of chicken breeders across the country to produce the ultimate breed of meat bird—in other words, one that could provide the most meat in the shortest time and for the lowest feed cost (see page 13).

Throughout the world, poultry started to be used as an alternative source of protein. Hybridization through careful selective crossbreeding, coupled with a better understanding of the biology and genetics of poultry, meant that breeding programs could be constructed to pursue two discrete results: egg-laying "machines" and fast-growing meat sources. The heritage breeds were soon forgotten in the pursuit of cheap food, and it was left to the exhibition poultry breeders to maintain the gene pools.

Today, science and agriculture continue to work hand in hand with the supply chain to meet the ever-increasing demand for poultry-based products. More and more hybrids were, and still are, being developed to meet specific consumer requirements, and in less than 100 years we have shifted from hen fever to a time when chicken meat and eggs are now staples of everyday diets worldwide. It's a revolution in agricultural practice driven onward by consumer demand, and now almost forgotten in its wake sits its bedrock, the breeds of the show bench whose history now puts their much-needed conservation in context.

CROSSBREEDS & HYBRIDS

When venturing into chicken-keeping for the first time, and particularly when the objective is to have a good supply of fresh eggs for the kitchen, it is likely you will encounter people saying, "Get some hybrids. They're an excellent beginner's bird and filled with hybrid vigor, unlike the pure breeds." But be careful before taking the plunge on this count, as the commentary might not be backed by the science. Advocates of this approach may be inadvertently basing their claim on heterosis, or "hybrid vigor" as it's more commonly known, and consequently confusing the concepts of crossbreed and hybrid.

A crossbreed is generally any offspring that is the product of a mating between a cock and a hen of two different breeds. It is usually the product of an "accidental" mating of the two different breeds rather than a premeditated mating aimed at introducing a genetic trait into a comprehensive long-term breeding program.

A hybrid, on the other hand, is the offspring produced when two different breeds or strains of chicken are intentionally crossed. These chickens themselves may not be pure breeds but a generation of hybrids. Hybrid breeding tends to be performed with a specific end game in mind; this could be for the production of a fast-growing bird for the table or an improved egg-layer. As such, commercial broilers and egg-layers are hybrids. They are fast-maturing and short-lived, with broilers rarely exceeding 10 weeks of age and egg-layers usually depleted (destroyed) once they have fulfilled one year of egg production. In reality, the term "hybrid" generally refers to all strains of chicken developed after 1950. As such, they have a very specific commercial application and turnover that is not always suited to the small-scale or hobby keeper.

Left
Hybrids have pushed many heritage breeds to the point of extinction.

Rehoming Rescues

When commercial laying hens come to the end of service (generally when they are around 18 months of age), the farm usually removes them and replaces them with new pullets that are approaching point of lay (POL). In most circumstances, the depleted stock are destroyed, but there are organizations that "rescue" the hens and then offer them to people to rehome.

The concept of taking a beast from a life within a factory farm and giving it the chance to live out a more natural existence could be said to be admirable, fulfilling, and compassionate, and in many respects it is. However, before taking on rescue hens, it is important to understand the situation fully.

The breeds used within commercial food production are not like other breeds of chicken. For a start, they don't have names like Orpington or Leghorn, but more often names that resemble car makes and models, like Hybrid V56N. They are often produced by genetics companies that provide detailed reports referring to the "product performance." These chickens are specifically created to serve a purpose and are frequently accompanied by a product sheet rather than a breed profile.

These "products" are designed with a fixed shelf life in mind. At no point is there any consideration, or need for consideration, beyond the point of depletion. The design brief for them would read something along the lines of "Reach maturity as rapidly as possible, lay as many consistently sized eggs as possible, and do it with the most efficient feed conversion ratio within the first year of laying. Replace when peak performance drops."

Anyone who has rehomed commercial laying hens will have experienced the incredible transformation they go through, and no doubt this inspires others to do the same. However, before you do so, it is important to have a full knowledge of the potential problems. Commercial breeds are designed to be egg-laying machines, and even after their rate of laying drops below commercially acceptable numbers, they will continue to lay at such a rate that they run the very real risk of laying themselves to death. Additionally, in later life, they can develop all manner of health problems.

There is much to be gained from rehoming these "products" and letting them be chickens. Yet it is important to be mindful that the accelerations in poultry science mean the creation of egg-laying hens designed for indiscriminate depletion has become acceptable. Anyone embarking on giving such hens a new lease of life must be prepared to act humanely if necessary.

Ethical Eggs

The market share of free-range eggs has increased significantly in recent years because larger numbers of consumers have become more concerned about where their food comes from and, in particular, the ethics of egg production. This groundswell has resulted in a much wider understanding of the production chain, but only in terms of the manner in which the laying hens are kept and the environment in which the eggs are being produced.

The wider picture is more complex, because for every hybrid laying hen that hatches, there is a hybrid male—in other words, half of all chicks that hatch are male and half are female. What many people don't realize is that these male chicks are invariably slaughtered on day one because they serve no commercial purpose.

No matter how ethically sound the husbandry system in which the resulting pullets are kept, the reality is that hybrid male chicks from those strains designed for egg laying rarely survive beyond hatch day. This is because the egg-laying hybrid is designed specifically for egg production and so does not fatten up, meaning that the males do not make an economically viable source of meat. The issue of what to do with the male chicks from these hybrid strains is not wholly ignored by those responsible for their future, with some significant research being applied to genetically modifying the chickens further. An example of this currently involves the potential of adding genes from jellyfish into the hybrid strain. This makes it possible to identify the female embryos within the shell as the introduced gene causes them to fluoresce under UV light. The advantage of this is that the males can be disposed of prior to hatching, thus not only removing the issue of destroying millions of male chicks but also saving both money and the greenhouse gas-emitting resources expended in incubating the male eggs to hatching point.

If it's possible to put aside the ethics of killing the male embryos and any strength of opinion regarding genetic modification, this could be valid solution to the problem. However, there are potentially better solutions. For example, a major commercial poultry breeder is trying to develop a breed of chicken whose females not only lay well but whose male offspring have good table properties—in other words, tapping into the real value of dual-purpose breeds. To a small farmer, the dual-purpose breed is not only economically sensible but also ethically sound, where nothing is wasted and everything serves a purpose. It is worth considering this wider picture when deciding on which chickens you intend to purchase.

Buying Your Chickens

One of the best times to pick up new POL stock, both in terms of bird health and from a price perspective, is late summer onward. This may seem an odd time to buy, with most people probably thinking that spring would be the best, but a bird that is 20–25 weeks old in the summer will have hatched in early spring and will hopefully have been allowed to grow outdoors during the warm summer months. It's also very possible that the bird will come into lay before the onset of winter and subsequently lay throughout the winter, when older birds are resting or molting. In the spring, any bird at 20–25 weeks of age will have spent significantly longer indoors, possibly under artificial heat/light for a larger part of its growing period, and this can impact the overall quality of the livestock.

In late summer, many breeders will have finished selecting their stock for the next season and exhibition breeders will have picked out their show team, so there will be a lot of surplus stock on the market. Much of the stock will be high quality for the reasons given above, making the prices more competitive than earlier in the season. This provides the clever buyer with a good chance of grabbing a bargain.

STOCK TYPES

Frequently, the stock on sale is described as either show-, breeder-, or pet-quality. These categories are defined as follows.

Show-Quality Stock

An example of a bird that is deemed as being of show or exhibition quality is one that exhibits all the visual attributes required for the standard of that particular breed and plumage type, and that, at the right time of year and when prepared for an exhibition, would stand a chance of being placed and receiving an award. The key points here are visual appearance and standards. These birds should meet the requirements of the show bench, even if they do not necessarily meet the original requirements or intentions in the development of the breed. This is a point worth bearing in mind when setting out to purchase a pure breed, as show-quality birds may serve you well in a show but not in the kitchen.

Above
Poultry auctions provide a chance to grab a bargain.

Breeder-Quality Stock

These birds will have some defect within their features that prohibits them from taking honors at a show, but they do have the genetic makeup and potential to be used to breed a show-winning bird. It is a common misconception that two show winners, when bred together, will automatically produce many more show winners. They don't and, in fact, rarely will.

Pet-Quality Stock

Usually these are purebread chickens that are sub-show standard and shouldn't be used as part of a pure-breed breeding program. It doesn't make them any less of a chicken, and, despite the name, it also doesn't mean they are necessarily pet-like in their behavior or docile in their temperament. You still need to make sure you select the right sort of breed for your requirements when choosing pet-quality stock.

BUYING ONLINE

The actual sourcing of your poultry can seem a little difficult at first—after all, you don't often see livestock stores in the local shopping mall where you can stop in and check out what they have to offer. One alternative is the Internet: when it comes to buying poultry, it's one big storefront offering access to practically every breed of chicken and type of poultry paraphernalia available.

With the breadth of options available on the Internet, it's easy to see the advantages in terms of access to possible purchases. Images, text, and interaction with the sellers can certainly deliver a level of confidence in ensuring that you are likely to make a good purchase, but there are pitfalls to watch out for. It's important to remember that all you see and read is "virtual," and even when you've made your decision on which seller to go with and what to buy, you need to apply the same purchasing rigor as if you were buying at a live auction or face to face from a breeder.

There are a number of ways in which the Internet makes eggs and birds available to potential buyers. This can be via auction websites, through trading websites, through individual breeder websites, and through club sites and forums.

Auction Websites

Auction websites either sell hatching eggs or live birds, or some do both. Most don't require you to register in order to peruse what's for auction, but all do require you to register if you want to buy or sell. Most, if not all, keep your identity private (at least until the payment stage) and rely on the use of "tags" or nicknames to identify individual users.

The advantage of auction websites really comes into its own when sourcing hatching eggs. These can be sent by mail or courier, which, assuming they are packed correctly, will have no adverse effect on their viability. With careful bidding you can often pick up a bargain, but by the same measure you can waste money on duds.

Livestock auctions are a little different because special arrangements must be made to ship live animals. Sometimes sellers will allow you to visit and see the stock before bidding, something that can certainly help in making a decision to purchase.

Trading Websites

Some websites act as a directory enabling people either to advertise single items for sale or to promote their small businesses (Facebook is also increasingly being used by groups as a method advertising items for sale). Like the auction websites, these can be very useful when it comes to sourcing stock and comparing as many outlets as possible in one place.

Unlike the auction websites, only the seller needs to register in order to place an advertisement. While registering as a buyer will enable you to contact the seller via a messaging system on the site, most of the ads carry direct contact numbers, which negates the need for buyers to register and allows them to get in direct contact with the seller.

Breeder Websites

Some poultry breeders set up their own websites to advertise their business and the items they have for sale, including the breeds they have available and the stock they are carrying. More often than not, contact details are provided, enabling you to get in touch with the breeder directly, either by phone or email, or by visiting them in person. These sites can, however, be more difficult to find on the Internet because they don't usually turn up at the top of results lists in search engines unless they are affiliated with a larger organization or have exceptionally large numbers of visitors.

Club Sites & Forums

Poultry and self-sufficiency/lifestyle-based websites are also worth checking out. These often have forums where visitors can register and take part in conversations with other members on the site. Some poultry club and magazine websites also use forum technology. Aside from the fact that these forums are a great way to find like-minded people and share knowledge and experience in much the same way as would a physical club, they also usually have "for sale" and "wanted" sections where members can post items. These can be very useful for locating stock or requesting sources for stock, and may also often result in a link back to one of the other types of website already mentioned.

QUESTIONS TO ASK THE SELLER

When you have found a seller who has the stock you are looking for, phone, email, or visit them and ask the following questions. Remember that if you are unhappy with any of the answers they give, you are under no obligation to buy. Be patient and be prepared to wait for stock. Good breeders are often small-scale outfits and as such are likely to have a waiting list.

- Ask to see the parent birds and where the young birds were reared. Good breeders will be happy to show you their setup, but do respect their biosecurity measures.
- Ask if and when the birds were wormed and/or vaccinated and what with. Good breeders will keep records and will provide you with details.
- Ask whether there is a return policy. Breeders will always tell you whether the stock they are selling is sexed or not. Mistakes do happen, but a good breeder will exchange or provide a refund in such an event.
- Ask whether there is a guarantee period. Good breeders will allow a period of around two weeks when the birds can be returned for a refund if there is a health problem.
- Always ask to hold the birds and check them over first before committing to buy (see list that follows). Good breeders will be more than happy to have their chickens checked by a buyer before they are sold.

BASIC CHECKS BEFORE YOU BUY

When you are handling a bird prior to purchase, perform the following basic checks:

- Part the feathers, particularly around the base of the tail and around the vent, and check for signs of parasites such as eggs on the feather shafts.
- Rub your hand over the legs of the bird. The scales should be flat against the shanks and not raised or rough, which could be a sign of scaly leg mite (see page 171).
- Hold the bird up, place the side of the breast to your ear, and listen to its breathing. Any wheezing or rattling could be a sign of a respiratory disorder.
- While holding the bird up, take the opportunity to sniff its breath. There should be little or no smell. Any stronger odor could be a sign of a digestive or crop-related problem (see pages 167–8).
- Feel the bird's crop (see page 166). Healthy birds will be eating and therefore have food in their crop—this should feel like a slightly soft ball. An empty crop or one that feels either solid or squishy are signs of an unhealthy bird.
- Feel the bird's general condition—it should be firm, not fat or skinny.
- Check that the eyes are bright and the comb is the correct color and that the bird is generally alert and not lethargic.

Again, if you have any concerns, remember that you are under no obligation to buy your chickens from the seller and are free to look elsewhere.

Home to Roost

Once you have purchased your chickens, make sure their house is ready and the fencing is secure prior to picking them up. Open the ventilation in the house if it's not open already and fill the feeders. If you have an outside feeder, either place it inside the coop if space permits or place a small bowl of feed in the coop instead. Also, put a small waterer in the house temporarily. Waterers are generally best kept out of the living quarters because the chickens will invariably knock them over.

Before setting off to collect your chickens, check whether the seller will supply transport boxes. If not, you will need to take along either a sturdy cardboard box or a pet transporter of an appropriate size. Food and water are not essential for the occupants on the journey unless it is going to be particularly long, but try to make the collection time as late in the day as possible to avoid any high temperatures. As you put each chicken into its box, give it a final check to make sure it appears healthy.

When you get the chickens home, check the birds over once again and then place them in the coop. Don't let them out into the run, but instead leave them indoors with the food and water so that they can settle in. They will roost in the house overnight, and hopefully when you let them out into the run the following morning, they will realize the coop is home and where they should roost. After a couple of days, the waterer can be placed outdoors, along with the feeder if you intend to have it outside.

If you intend to allow your birds to roam free-range, it is worth initially keeping them within a more confined area. This will enable them to familiarize themselves with the immediate environment, with you, and with any other regular presences such as other animals and passing traffic. After about a week, remove the temporary enclosure, and your chickens should be ready to roam.

If you already have chickens, then keep the new stock separate from them for the first two weeks. This is a sensible biosecurity measure and decreases the risk that any disease or parasite may be transferred. After this time, you will be ready to integrate the newcomers into the flock. This can result in a bit of aggression between the new and old as the pecking order (see box on opposite page) is established. Such aggression can be kept to a minimum by introducing the new birds to the coop a few hours after the existing flock has gone to roost. This will at least mean the new birds are able to rest for a night before they attract the attention of the old birds.

In the morning, ensure that more than one waterer and feeder are available, as again this will help reduce aggressive encounters between the two groups of chickens as they take food and water. No matter what measures you take, it's likely that there will be some aggressive behavior between individuals from each of the groups as the pecking order is established. This is only natural and in most cases the new birds will be subordinate to the existing flock until a more familiar social hierarchy can be established. Occasionally, though, aggressive behavior can lead to outright bullying. This is when there is constant hostility by one bird toward a subordinate bird. In these instances the bully should be removed from the flock and placed in a quarantine cage for a week until the subordinate has the chance to settle. Do not remove the bullied bird (unless you relocate it permanently), as its reintroduction to the flock will invariably result in a return of the original bullying behavior.

Aggression toward people

While aggression between chickens is part of their normal behavior, aggression toward people should not simply be dismissed as dominance or bad temper. In this case, the bigger picture should be considered and, in particular, the behavior of the keeper.

First, if there is a cock in the flock, ask yourself if its hens are persistently being picked up? If it is an all-female flock, is the underdog favored and given more attention and treats than the others? Such apparently insignificant interventions have an impact on the social structure of the flock and could give rise to competition and aggression.

Pecking order

This is the social hierarchy of a flock of chickens and can develop when the birds are as young as six weeks of age. It will determine which birds eat first, which will have the best dust-bathing spot, which will have the highest roost point, and so on. It delineates the dominant from the subordinate, and is usually established through a degree of aggression (pecking). It does, however, create cohesion within the flock structure, which, while competitive in its concept, is also there to minimize conflict.

Pecking orders exist in all flocks, be they single or mixed-sex. In general, if the flock contains a single cock, he will be at the top of the pecking order, followed by a hierarchy of hens. If the flock contains a number of males and females, then the order tends to be roosters or cocks, hens, cockerels, and then pullets, with the members of each group having their own order of rank.

Many factors can influence a bird's position in the pecking order, and it doesn't always come down to size or physical fitness. For example, a bird with a large single comb, such as an Ancona, will often find itself higher up the hierarchy than an Araucana, which has only a pea comb, despite the Araucana being bigger in size.

Managing Flock Structure

Aggressive behavior between chickens is, in effect, a form of negotiation. Human society has developed negotiation to the point that, in most instances, it is performed using verbal communication, avoiding conflict, although in many cases it will still carry a level of posturing.

Within chicken society, negotiation tends to start with posturing, and, if that doesn't resolve the problem, conflict will follow, be it a sharp peck, a kick out with claws and spurs, or a full-on fight.

While chickens will invariably establish their own social structure, it is possible for the keeper to assist in keeping it stable and minimizing stress, tension, and unnecessary conflict. This can be done by:

• Providing sufficient feeders and waterers so that more birds have the opportunity to feed or drink at the same time.
• Ensuring that there are always places to hide for the lower-ranking birds.
• Providing sufficient space (see page 111).
• Minimizing disruption to the pecking order by not adding or removing birds too frequently.

Right
Flock cohesion and structure play an important part in chicken "society." It is important that the keeper recognizes this.

Care Routines

You will need to create a regular routine of tasks for the care of your chickens each day, week, month, and year (see table on page 138). While this regime will cover most of your chickens' requirements, there are particular seasonal considerations during periods of extreme summer heat or cold and wet winter weather.

In both circumstances it is important that shelter from the elements is provided—shelter from the heat of the sun or shelter from biting cold winds and rain. This can be as simple as leaning a fence panel against the side of the coop. Water is another key consideration. Chickens drink quite a lot of water no matter the season. Ensuring that the water you give them is fresh, cool, and free from algal growth in the summer is as important as ensuring that it isn't frozen solid during the winter.

If the ground gets muddy during the rainy season, provide raised areas for your chickens so they can get themselves out of the mud. And if it snows, clear areas for them—snowfalls deeper than 4 inches (10 cm) are generally considered impassable by chickens.

How to clip a wing

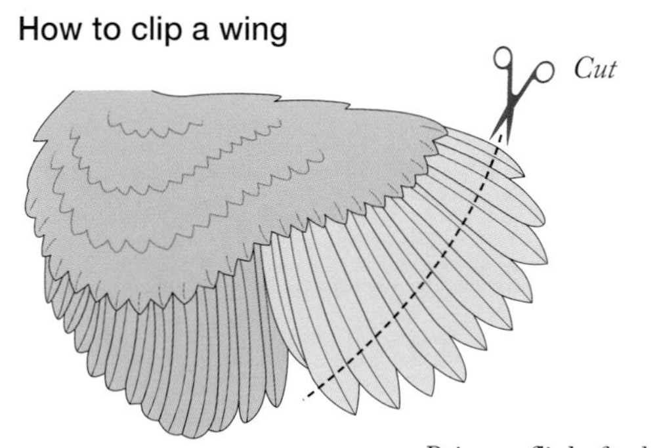

Wing-clipping

While it is not a requirement to clip the wings of your chickens, you may need to do so if you have a bird (or birds) that persistently clears boundaries and heads off to the neighbor's backyard or your vegetable plot. Some breeds of chicken are reasonable flyers over a short distance, while others are very good jumpers. Despite clipping, however, some birds will still manage to clear a fence. Wing-clipping is not, therefore, a fix-all, and some sort of netting over the run might be a better alternative.

Wing-clipping is easiest if you have an extra pair of hands to hold the bird. With sturdy scissors, cut off the primary feathers on one wing only. This serves to unbalance the bird and reduce the height it can gain. The primary feathers are not seen when the bird's wing is closed, so cutting them doesn't affect the look of a resting bird.

Wing-clipping should be done when the feathers are fully grown and only when the bird is mature. The feathers will remain cut until the bird molts, at which point new primary feathers will grow. Once these are fully grown, you can clip them again, but hopefully by that time the bird will have gotten out of the habit of ranging too far.

Regular Care Regimes

Time period	Specific tasks	General tasks
Daily	Provide fresh water Check feeders and refill if required Check nestboxes for eggs One hour before roost, provide a scratch feed of corn or wheat, check for eggs again, and refill waterers if required	Watch your birds whenever you get a chance so that you come to understand their character and behavior—this is livestock husbandry "by eye."
Weekly	Muck out the house Wash, disinfect, and dry waterers and feeders Refresh the dust-bath if using a purpose-built design Move chicken tractors to fresh ground	Give the birds a physical once-over to check their general health and ensure that they are free from external parasites.
Monthly	Give the henhouse a thorough, deep clean Dust birds with antilice powder Check and clip toenails and spurs if needed Refresh ground litter if the birds are kept in a fixed run	Give the henhouse a once-over to check for anything in need of maintenance or repair.
Annually	If the house is wooden, treat it with a natural wood preserver Worm the flock twice yearly (spring and fall)	Throughout, take every opportunity to handle the birds and check their well-being.

Spur-clipping

As male chickens age, their spurs can grow quite large. To help avoid the injury the spurs inflict on hens during mating, and to prevent the spurs from becoming snagged or growing inward toward the leg, it is a good idea to clip them and file them to a blunt end.

Large dog nail clippers make an ideal tool to trim the spurs, although care should be taken not to take too much off. There is a blood vessel running down the center of the spur—usually about three-quarters of the way along it—that should not be cut, as doing so will cause bleeding that can be difficult to stem.

Biosecurity

Biosecurity is a buzzword in livestock farming, but it is one that backyard poultry-keepers need to be aware of. No matter how small your flock of birds may be, good biosecurity practices should be followed, not only to minimize the risk of disease transfer within your own poultry, but also transfer to other people's birds. Below are a few commonsense biosecurity measures to build into your daily routine.

- Keep poultry feed under cover to deter the attentions of wild birds. Ensure that water is always fresh, and clean out waterers at least twice a week, if not more.
- Replace any water that becomes soiled with droppings.
- Quarantine any stock that has been off site (such as to a poultry show) for at least seven days.
- Quarantine new stock for at least two weeks before bringing the birds into contact with existing stock.
- Clean your clothes and boots after visiting another poultry establishment, show, or sale.
- If you have more than one pen of birds, consider using a disinfectant boot wash.
- Don't share transportation crates or feeding equipment with other keepers.
- Always disinfect transportation crates before and after use.
- Wash your hands before and after handling poultry.
- Keep vermin such as rats under control.

Biosecurity is all about disease prevention. By following these simple precautions, you will go a long way in protecting your flock from infectious diseases.

Dealing with Unwelcome Visitors

No matter what country you live in, there will invariably be a number of local animals that are keen to make a meal out of your flock of chickens. Foxes, badgers, weasels, and even hedgehogs will break into a chicken run and attack chickens if they are hungry enough and if the security of the flock is not strong enough.

It is important therefore to consider the animals in your immediate environment and deploy security techniques accordingly—though keep in mind that some may be stronger than you (bears) or have opposable thumbs (raccoons)! Whichever method you settle on, the following basic principles will help reduce the likelihood of an attack:

• Don't let the flock out too early in the morning, especially into an unsecured area, as nocturnal predators might still be out. If other animals in the area are keeping quiet, it's quite possible a predator might be lying in wait. Conversely, noisy wild birds might be a sign that a predator is near.
• Keep boundaries well secured. Predators will look to exploit any weakness, be that a short circuit in an electric fence or a hole dug by a rabbit. It might take weeks for them to find a weakness, but they will find one. You must constantly be on your guard.
• Walk around your flock at irregular times. A pattern to movements is no different than a weakness in your fence, and a predator will quickly learn to exploit it.
• When the birds go to roost, be there to close the door; in fact, be there 15 minutes beforehand to check that all the birds are in and there are no stragglers.
• Be aware that many nighttime predators are totally nocturnal and may also feed during the day.

PETS
Pet cats and dogs can be a concern to a chicken-keeper, and it is always worth familiarizing them with your chickens where possible. Cats will invariably leave large, full-grown fowl alone, though bantams and young growers can be at risk of being attacked and killed. Some dog breeds, such as Collies, mix very well with chickens, happily coexisting alongside them, but other breeds, such as terriers, can pose a risk. The chicken-keeper will need to train the dog to keep away from the chickens, though any dog that expresses an unhealthy interest in a flock should never be left alone with it.

If you do not have a dog but have visitors who bring their pet along, politely point out that while their dog might not be bothered by the chickens, they will be bothered by him. If they are not familiar with dogs, most chickens are intelligent enough to view them as a predator threat.

RATS

Rats are fairly ubiquitous creatures and can be an issue for the chicken-keeper year-round. They are a particular nuisance in winter, when the cold weather sees them struggling for food and shelter and potentially seeking out a cozy chicken run until the weather improves.

Having rats in your chicken run is not a sign of poor husbandry, but it is poor husbandry not to deal with the problem once identified. Rats need three things—food, water, and shelter—so by removing at least one of these requirements, there's a good chance you will remove the problem. Don't leave feeders outside; either bring them indoors or put them inside the chicken coop at night. Empty the waterers each night and refill them in the morning. Finally, raise the coop off the ground, if possible by at least 6–8 inches (15–20 cm), to prevent rats from nesting underneath it (this has the bonus of providing an outdoor shelter or new dust-bath location for your chickens).

Rats present a limited predator risk and are only likely to attack young growing birds or very small bantams. There is also anecdotal evidence that rats will be attracted to the coop in order to steal eggs. There is the possibility that small bantam eggs could be taken by a rat, though eggs from large fowl, especially when located in a nestbox, are too awkward for a rat to move. If a broken, half-eaten egg is found in the nestbox, the finger of suspicion should be pointed elsewhere first.

AVIAN PREDATORS

Crows are very intelligent birds and are not afraid to enter a chicken coop and help themselves to any food, particularly the eggs inside. The will usually try to consume the egg in situ, especially if access to and from the coop is restricted. This can result in a broken, half-eaten egg lying around, which could in turn attract the unwelcome attention of a chicken that may then pick up the difficult-to-break habit of egg eating (see page 150).

To reduce the likelihood of crows or other corvids entering the henhouse, simply place a loose material screen cut into strips over the door of the pop-hole or entrance to the nestbox once your chickens have gone in to roost. Come the morning, the chickens will work out how to brush past the material and will enter and exit the coop as if it had always been there. If the sheet is put over the nestbox, the chickens will appreciate the additional darkness and privacy it provides when they are laying. Crows, on the other hand, are unlikely to enter the coop or the nestbox because they seem to avoid doing so unless they can see a clear exit.

Small backyard birds are unlikely to present a direct danger to your chickens, but they will steal food and can carry pests, such as lice and mites, into the coop and run area. Using small-gauge netting or wire around and over the run will keep them out. If your chickens are free-range, however, contact between them and wild backyard birds is inevitable at some point, although food theft can be reduced by making use of a treadle feeder (see page 118).

Larger wild birds, such as birds of prey, have been known to attack and take chickens for food. Roof nets can be added to the runs of flocks kept within a confined space to stop any chance of attack, but free-ranging chickens will be at risk if large birds of prey are present in your area.

CHAPTER NINE

Laying & Brooding

Establishing a flock of laying hens can be extremely fulfilling, and if the first birds purchased are at point of lay, then the first egg will be an eagerly awaited event. There is little to compare with the taste and texture of a fresh egg, especially if only moments earlier it was collected from the nestbox of the coop in your backyard or small farm. Quite frequently, people find their flock is producing more eggs than can be consumed by the household. Six regular hens will each produce around 200 eggs a year (and that's an underestimate for some hens).

Laying Cycle

The laying cycle is the number of consecutive days that a hen will lay an egg and can vary from around 14 days to close to a year at the commercial end of the scale. It is a myth to believe that once a hen starts to lay, she will do so without a break until her next molt. The laying cycle can also vary according to the breed, with a pure dual-purpose hen perhaps managing 30–40 days before breaking for a day or two, and a good laying breed like the Leghorn managing up to 80 days.

The rate of lay refers to the number of eggs a bird will lay in a calendar year. It is very variable and depends primarily on the breed and strain of chicken. Some laying strains of Leghorn, for example, can lay up to 300 eggs in a year, while an exhibition counterpart of the same breed may achieve only double figures.

The period of lay is the time lapse between the first and last eggs of the season. This time lapse can (allowing for breaks that would be expected due to the laying cycle of the bird) be dependent upon many environmental factors, the stresses these place on the pullet, and the breed of chicken. However, a young pullet or young hen will usually lay from early spring through to late fall.

Period of lay may also refer to the longevity or number of years a hen will lay. Again, this can depend on the breed and strain of a chicken, but it can also be influenced by the conditions in which the bird is kept. If the hen is simply exposed to the normal environmental conditions (temperature and light) dictated by the seasons of the year, she will lay between spring and fall each year, and hence could go on to provide eggs (albeit it at a lower volume as she ages) for up to 10 years and possibly beyond.

If, however, the chicken is subject to a managed environment where optimum laying conditions are provided throughout the year, she may lay (albeit in higher numbers) for only three years, by which time she could have exhausted herself and her supply of eggs.

Light & laying

While hens need at least 12 hours a day as a stimulus to lay eggs (see pages 20–1), the optimum is 15 hours of light a day. When day length drops below 15 hours in late summer, you can simulate optimum conditions with the gradual introduction of controlled artificial lighting. This will encourage your chickens to continue to lay beyond the point when they would naturally slow down and stop, usually prior to their annual molt. Light intensity also plays a part in the stimulus of laying, and incorrect lighting can have a negative effect on a chicken's ability to rest and develop immunity.

Selecting & Selling Eggs

Even accounting for differing laying cycles, a flock of six chickens can produce around 30 eggs a week. Unless every member of your household eats an egg a day for most of the week, this will result in a surplus. It is possible to preserve eggs for those egg-free periods of the year by pickling or freezing them (see box), but the benefit of a home-produced egg is its freshness, so why not sell them? The income will go some way to help fund the upkeep of your hens and may even provide a small profit.

Preserving eggs

Pickling eggs

Simply hardboil the eggs, peel them, pack them into a pickling jar with a tight-fitting lid, and add pickling vinegar and any spices or herbs you might like. The jar should be stored in the refrigerator for five days and then put in a cool, dark cupboard. The eggs can be stored for up to four months.

Freezing eggs

Bake the eggs in muffin tins for 30 minutes, leave them to cool, then double-wrap them with plastic wrap and put them in the freezer. They will last up to six months and can simply be defrosted and reheated when you want to use them. Alternatively, scramble the raw eggs and place portions in freezer bags. The scrambled eggs will last more than months and simply need to be defrosted in a refrigerator prior to reheating.

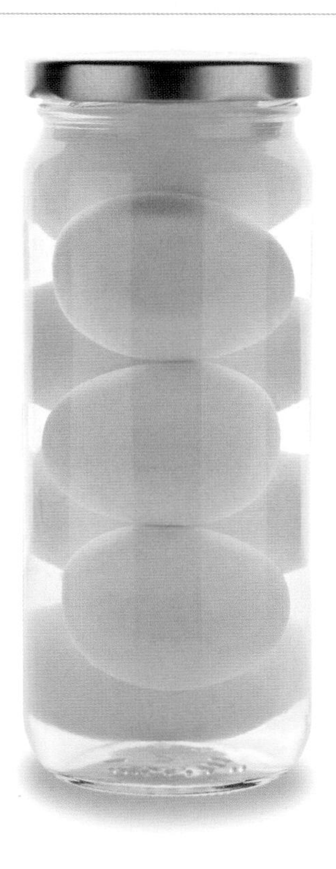

Before you do so, it is important to check the law regarding the sale of eggs in your town, county, and state as well as any federal laws. Your county extension agent or your state's agriculture department will be a good source of information about the laws that apply to small-scale egg producers and sellers. Selling eggs in this way bypasses the regulations of mass production but in some areas does carry restrictions that need to be adhered to.

Being a good egg

Many backyarders and small-scale farmers keep chicken flocks that produce more eggs than are needed by the household, so they sell their surplus eggs. Many who don't sell with profit necessarily in mind do sell with the intention of at least covering the costs of production, and this is where the price of the product can be pivotal.

There is little point in pricing your eggs higher than an equivalent product sold in the local shop or supermarket unless you have some other unique selling point. By the same measure, if you are in a remote location with little passing trade, bargain prices may be needed in order to attract some customers.

Be mindful though not to devalue your hens' production by practically giving the eggs away, and do not thoughtlessly undercut the sellers down the road who have given their pricing some consideration. Be a good egg and charge a fair price for a fair product. That way you won't lose customers, and in fact you might even gain some.

Cleaning eggs

When eggs are laid, they are slightly wet to the touch. This soon dries out, but there is a chance that nesting material such as wood shavings will adhere to the egg. This should brush off easily enough, and if nestboxes are managed correctly, and are of the correct design, this will be the only "dirt" in need of removing. Any egg that is badly fouled with mud or droppings should be disposed of, as it is likely bacteria will be present, which could be transferred through handling in the kitchen.

In commercial situations, eggs for eating are often washed prior to sale; however, this requires precision because it will remove the surface bloom, which helps prevent any bacteria present on the eggshell from entering the egg. In a small-farm situation, if cleaning is required, it should be done only lightly, with a dry nylon scouring pad. If, however, the eggs are not intended for consumption but required for incubation, then they can be washed. However, only the most valuable eggs should be washed, as doing so increases the risk of bacteria entering the egg, killing the embryo within, and potentially causing the egg to explode in the incubator. This in turn could damage all the other eggs being incubated, as well as the incubator itself.

Washing should always take place using an approved egg sanitizer and according to the manufacturer's instructions. This will usually involve immersing the eggs in a slightly warmer solution in order that fluids (and bacteria) are not drawn inside.

EGG SALE DOS & DON'TS

• Do make sure you research prices locally. Selling direct (and especially when not driven by profit) means it's easy to undercut the supermarket prices, as you don't have their overheads, but do take into consideration the costs of production. Try to price the eggs so that the income makes a contribution to the cost of keeping the hens; they will still be an attractive alternative compared to the higher-priced eggs sold by larger retailers.

• Do candle the eggs to make sure they are not cracked, and then store them in a cool location out of direct sunlight until they are sold.

• Do label your egg boxes with a "best before" date. Eggs will last for up to four weeks from the date of laying if stored correctly, so make sure you have some system to order your produce by date and ensure your customers know the best before date when they buy. Check with your local authorities whether there are any other labeling requirements for egg boxes, such as the producer's address or the need to refrigerate after purchase. It's more than likely you will be selling eggs that are only a few days old, which puts them well ahead on freshness versus store-bought eggs, but it makes good business sense to ensure your customers are aware of the best way to store their eggs and their likely shelf life.

• Don't grade your eggs by size. Do box them up according to the size/type of the chicken that laid them—for example, large fowl or bantam—but avoid stating if the eggs themselves are large, medium, or small, as these are frequently commercial terms that require the egg to conform to a regulatory size.

• Don't try to sell the eggs as free-range or organic unless you have official certification. These terms again tend to be industry standards and require registration in order to sell eggs as such. Instead, use some imagination in your marketing—"Backyard grown, friendly and free to roam" might be a suitable alternative description.

• Don't forget to give some thought to how you intend to market your eggs. It's surprising how many people prefer to buy their produce directly from the farm, where freshness can be counted in hours or days and not weeks (as at the supermarket), and food miles on fingers as opposed to a calculator.

• Most of all, keep a record of how much income your egg sales are bringing into the household. It may not be a huge quantity of cash, but understanding your expenses and income is an important part of managing your laying hens, especially if you intend to keep them as a self-funding supply of food for the home or farm.

Egg Issues

Usually, chickens will go about producing eggs without much trouble. If the husbandry is good, the feed and forage are of quality, and the stock is from good bloodlines, then issues with the eggs produced should be few. Eggs sold commercially are graded for quality, as described on page 37. While small-scale producers are not required to grade their eggs, it is good to be aware of the more common issues.

MISSHAPEN EGGS

The precise shape of an egg is defined in the isthmus during the egg's formation (see page 23). The general shape of the egg—pointed at one end, and rounded at the other—has evolved over time so that its surface disperses pressure. This means that when the egg is pressed from either end, it is remarkably strong compared with when pressed around the middle, where it breaks more easily so that the chick inside can hatch out. And when rolled, it will not roll straight like a ball but instead in a curve so that, given no obstacle and a level surface, it will return to its original resting place.

While commercially produced eggs all appear to be regimented in shape and size, this is primarily because they are graded before packing and are laid by genetically manufactured, hybrid, laying hens kept in their millions for the consumer market. Within a small flock of hens kept in a homestead or backyard, particularly if they are a mix of breeds and bloodlines, it will soon become apparent that each hen lays individually recognizable eggs in terms of their shape and size. While most will be the usual egg shape, some hens will lay almost round eggs and others that are toward the torpedo end of the scale. In addition, eggs may be "slab-sided," with a flat side, or "body-checked," with a bulge around their middle (see page 39).

TINY EGGS

Occasionally a chicken will lay a tiny yolkless egg. This is when the albumen and shell have formed around a small foreign body, such as part of the oviduct lining. These eggs are often referred to as "wind eggs," from a belief that they were created when a hen broke wind, or in some regions as "witch eggs" or "fairy eggs," because they were believed to have been placed in the nest by either a witch or a fairy. Another term used is "cock egg"—in other words, the feeble production of a male bird emulating the dynamic capabilities of the hen. They are not uncommon to find in the nestbox and usually occur at the beginning of a period of lay or toward the end of a laying season.

SOFT-SHELL EGGS

These are eggs that have only the soft inner and outer membranes in place. Although referred to as "soft shell," this is in fact incorrect as the egg is laid with no shell at all. They can occur in two circumstances: when the hen is rushed into laying her egg through stress or excitement, so that it is laid before the shell is formed; or when the hen has a deficiency in calcium or vitamin D within the diet (see page 163). Deficiencies of this nature can occur during warm weather when the hen is eating less and isn't getting sufficient calcium in her diet to produce the eggshell. It's not unusual for a hen to lay one or two soft shells in a season. However, if a hen consistently lays eggs like this, there could be an underlying viral condition that will need to be checked by a vet.

WRINKLED EGGS

Occasionally eggs with a rippled, wrinkled surface are laid, and in some cases they even appear to be creased. These tend to be laid by older hens and occur when the egg is not "plumped" sufficiently in the isthmus or when it is retained in the shell gland too long. They can be eaten, but they look unsightly, so they should be discarded from any egg collections intended for sale. If younger birds lay misshapen eggs, then it is possible there are other underlying causes that should be investigated by the keeper. Stress and overcrowding within the flock can contribute to imperfect eggshells, but if the environmental conditions are optimal, then a defective shell gland or disease such as infectious bronchitis could be the cause.

DOUBLE-YOLKERS

A double-yolk egg contains two yolks and is usually the result of two yolks being released from the ovary at the same time (see page 23). They occur more frequently in pullets than in established laying hens, although laying double-yolkers is often an inherited trait in heavy breeds of hen. The most yolks ever recorded in a single egg was nine.

Multiple-yolk eggs are not uncommon, and to accumulate hens that are capable of laying them was, and, to some extent, still is a prize worth having, particularly if you sell eggs from your home. That pleasure may not be shared by the hens, however, as double-yolkers are markedly larger than normal eggs.

Tip

In almost all breeds of chicken, the egg color is applied at the very end of the laying process, the egg having previously been white until that point (see page 24). As the egg leaves the hen, it is still slightly damp for a very short period of time, and at this point the coloring can be scratched off. If you intend to exhibit your eggs and therefore want them to have a solid color, free from any surface damage, then make sure the nestbox has no sharp edges or coarse bedding matter such as straw.

LASH

This is a meaty-looking deposit laid by a hen that may resemble an egg but feels rubbery to the touch—it's often described as looking like a chicken nugget. It occurs when part of the hen's reproductive system expires or collapses and breaks off, and then travels through the oviduct as if being laid. A lash usually occurs only in older hens and more frequently in commercial laying hens approaching the end of their natural laying life. It is unlikely the hen will lay an egg again after passing a lash.

FROZEN EGGS

Any eggs left uncollected during the winter can freeze if the temperature drops below 14°F (-10°C). The contents will start to expand and this will apply pressure against the shell, eventually causing it to crack. Frozen eggs are best discarded.

BAD EGGS

If a hen has been laying away from the nestbox and her clutch of eggs is found somewhere under a bush or in the corner of the barn, it can be difficult to ascertain just how long they have been there and whether they are safe to eat or have gone rotten. Fill a bowl with water and place each egg in the water in turn. If the egg sinks, its air sac is still quite small and therefore the egg is fairly fresh. If the egg floats, it has already lost a lot of moisture and the air sac will be large, providing it with the buoyancy it needs to float. Any floating eggs should be discarded as they are likely to be rotten.

If, on the other hand, the egg appears to move and creates small ripples as if it's vibrating, put it back where you found it, as it contains a chick due to hatch and you've just robbed a broody of her nest.

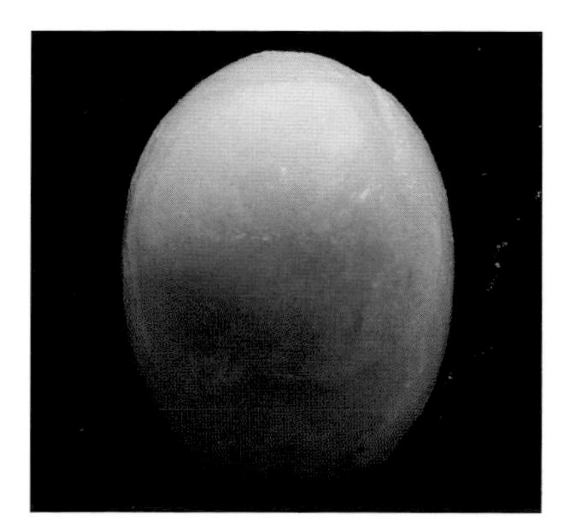

Page 148 left
The wrinkled egg isn't very appetizing to look at but frequently is OK to eat.

Page 148 middle
Often called a soft shell, it is in fact an egg with no shell at all.

Page 148 right
Double-yolkers should be prized as they make for an excellent addition to the breakfast plate.

Left
Scratch marks on the surface of the egg shows where the color was rubbed off before the egg dried.

Egg-eating

Chickens can develop some bad habits, and egg-eating is one particularly bad vice within a flock, primarily because it results in the loss of eggs but also because it is a habit that can be picked up by others and can be extremely difficult for the keeper to break. It's not unusual for a hen to investigate a broken egg that has been laid on the floor of the coop or cracked in the nestbox. It's also not unusual for the hen to eat the egg. However, if a hen is actively seeking out eggs to eat or is hanging around hens that are in lay, and even going to the extent of pecking the vent of the bird to encourage her to lay, then it is likely she has become a habitual egg-eater.

The first thing to do is to figure out what may have caused the habit initially. Are there sufficient nestboxes for all of the hens? Is the flock overcrowded? Is the diet of the flock suitably balanced? Are there any stress factors in the environment, such as overly bright artificial lighting? If any of these are obvious issues, fix the situation.

The next thing to do is to attempt to break the habit. Various techniques and remedies have been suggested over the years, most of which revolve around blowing an egg and filling it with chili paste or some other highly spiced concoction. The theory behind this is that if the hen decides to eat the egg, she will be so put off by the taste that she will stop eating eggs altogether. The sad truth, however, is that in situations where the habit is so ingrained, even the spiciest of eggs will not stop the hen. In these cases, the best course of action is remove the offender from the flock permanently.

FLAVORED EGGS

The freshness of an egg and the diet of the hen that laid it can influence its look and texture when it is cooked and also, to a degree, its flavor. Off-tasting eggs (assuming they are not old and stale) can occur if your chickens have eaten excessive quantities of garlic, fish oil, or strongly flavored fruit or vegetables. Because the eggshell is porous, eggs can also absorb strong smells from the surrounding environment that will influence their flavor. Creosote, fuel, and paints, along with more organic odors such as molds and natural fragrances, are capable of permeating the eggshell and affecting the flavor. The storage of the eggs produced by your chickens is therefore as important as the husbandry of the flock when it comes to egg flavor.

Below
The food consumed by a hen can impact the flavor of her eggs, giving them an odd taste.

Left
Broken eggs in the nestbox could simply be the result of a clumsy hen treading on them.

BROKEN EGGS

The presence of broken eggs in a nestbox is not always the sign that there is a bird with an egg-eating habit within the flock (see box on page 150) and could have another, more innocent, cause. Assuming you don't have a pest such as a crow entering the house and trying to steal the eggs, then it's quite possible an egg could simply have been broken by a less agile bird. Despite the fact that chickens appear quite nimble, some can be clumsy, and one such character may have stood on the egg and broken it—something that is more likely if the nesting material was kicked out by a previous occupant or if there were already a number of eggs in the nestbox.

One way to minimize losses of this nature is to use roll-away nestboxes. These are nestboxes with a slightly sloping floor, allowing an egg that has been laid to roll either forward or backward to a collection point. This removes the likelihood of the egg being crushed beneath the feet of the next hen to use the box.

BOUND EGG

Egg binding occurs when an egg gets stuck just inside the vent of the hen or pullet during laying. There can be a number of reasons this might happen: the egg may be too large to pass easily; the hen is unfit, ill, or overweight; or, in the case of a pullet, it is not yet fully matured and has not yet properly developed its egg-laying capability.

The condition can usually be identified by seeing the chicken repeatedly returning to the nestbox to lay but there being no egg in the nestbox when she comes out. Her distress will steadily increase and she may be seen bearing down in the run as she strains to lay the egg. If not dealt with, it can have serious consequences, as the egg is effectively blocking other biological functions.

The chicken will eventually tire and will not flee when approached. On handling the bird, the abdomen and vent area will feel swollen and hot to the touch. The chicken should be inspected to find out if there is an egg present. This can be established by a visual check of the vent area to see if the end of an egg is visible. If nothing is apparent, feel for a solid egg shape by gently handling the bird around the abdomen. Do not try to remove the egg by stretching the vent or inserting fingers into the vent, as you will risk injuring the bird and fracturing the egg. Instead, seek veterinary advice immediately.

The Broody Hen

If you have ever encountered a broody hen, then you will perhaps appreciate why they are deserving of their own section in this book. If, on the other hand, you have yet to encounter one, then by the end of this section you will hopefully have the knowledge and tools to manage when you do.

RECOGNIZING A BROODY HEN

"Broody" hens are hens that have decided to sit on a nest. Usually the nest will contain a clutch of eggs, but sometimes the hen will brood on an empty nest or one that contains dummy eggs (eggs manufactured from ceramic, which can be used to encourage broody behavior). The eggs in the nest do not need to have been laid by the hen, or even by hens of the same breed or, in fact, the same species. The absence of a cockerel in the flock will not necessarily stop a bird from becoming broody, but neither will the presence of one necessarily induce such behavior.

The hen will invariably have already stopped laying eggs by the time she exhibits broody behavior. She will also reach a level of blind determination such that, no matter what is put in the nest, be it a completely different set of eggs or even golf balls, she will attempt to incubate them. She may not be sitting in one of the nestboxes within the coop; many free-ranging birds furtively lay a clutch of eggs in a hidden location—such as under a hedge or in a shrubbery—over a number of days before simply disappearing from the flock, then return three weeks later with a small brood of chicks. When a hen goes broody, her behavior can change dramatically. What was once a mild-mannered bird that happily interacted with her keeper will now become an overtly aggressive creature. A broody hen will invariably be found sitting tight on her nest, and when approached by the keeper will puff up her feathers to make herself as large as possible and emit a "growl." If the keeper reaches towards her, the growl will become more intense, and eventually the hen will strike out with a peck or even momentarily leave the nest to attack. This behavior is very definite and should not be confused with a hen that is trying to lay an egg. Laying hens may express some concern at being interrupted but are more likely to move from the nest to a less vulnerable location rather than become aggressive.

If a broody hen is suspected within a flock but no hen is witnessed sitting, watch for the hen that appears to be irritable. A broody hen will be constantly in and out of the chicken coop, fluffing her feathers up and strutting around, clucking. She will behave with what could be described as a short temper toward other hens in the flock and the cockerel, if one is present.

Broody hens will take little food or drink and will preen infrequently, if at all. They generally remain broody for days or weeks, although birds of some breeds have been known to continue behaving broodily for many months. This can result in the bird's health being compromised and leaves the keeper with two courses of action: breaking her behavior pattern or allowing her to incubate fertile eggs.

AVOIDING BROODINESS

Some breeds of chicken are more susceptible to broodiness than others. However, all hens have the potential to go broody during the season. A broody hen will stop laying eggs while she is broody, and her behavior may also induce broodiness in other hens of the flock, which in turn will also cease laying eggs. This can be counterproductive when the purpose of keeping chickens is for egg production.

The precise triggers that cause broodiness are not easily identified, but to reduce its likelihood, the keeper should check for eggs and remove any that are laid as frequently as possible during the day, not just in the early morning and evening. This regular checking will not unduly disturb the laying hens, but it will prevent a clutch from building up that could entice a hen to sit.

BREAKING A BROODY HEN

Some hens that go broody can become quite determined, whether they are sitting on eggs or not. Physiologically, a hen is designed to brood eggs as part of its reproductive process, and the cycle of brooding lasts approximately 21 days from when they first start to incubate the eggs to hatching. During this time, the hen will rarely leave the nest other than briefly to feed and drink. Her preening, dust-bathing, and feather care will also be curtailed and she will lose weight, lose condition, and potentially pick up external parasites. If the eggs are fertile and hatch successfully, the hen will leave the nest and care for her brood and will also start to care for herself again. As such, the 21-day period is survivable for the hen and should not have any adverse impact on her long-term health. However, if a hen is left to brood with no prospect of a successful hatch, there is a significant risk she will remain broody for much longer than the 21-day period and in doing so will suffer damage to her health (aside from it inducing other hens in the flock to become broody). To avoid this, the keeper must either make use of her broodiness or break her broody habit.

Brooding is primarily the desire to nest; incubating and hatching actual eggs is secondary, as evidenced by the fact that a hen will brood fresh air if her internal switch is flicked on. In order to break broody behavior, it is necessary to place the hen in a comfortable yet clearly non-nesting place—a dog crate is ideal for the purpose. First, remove the plastic tray and place it on top of the cage as a roof (if the base is not removable, simply turn the cage upside down so the base now acts as a roof). Next, place the cage on two bricks in an exposed but shady outdoor position. Add food and water, and then place the hen in the cage. This can be done within the flock enclosure or away from the main group. The hen will protest, and this is when a battle of wills takes place between the keeper and the hen. The hen will not like being unable to nest and will want to return to her favored location. She will, however, settle. If, after a couple of days, her behavior seems more normal (pre-broody), she can be returned to the flock. If she becomes

The broody coop

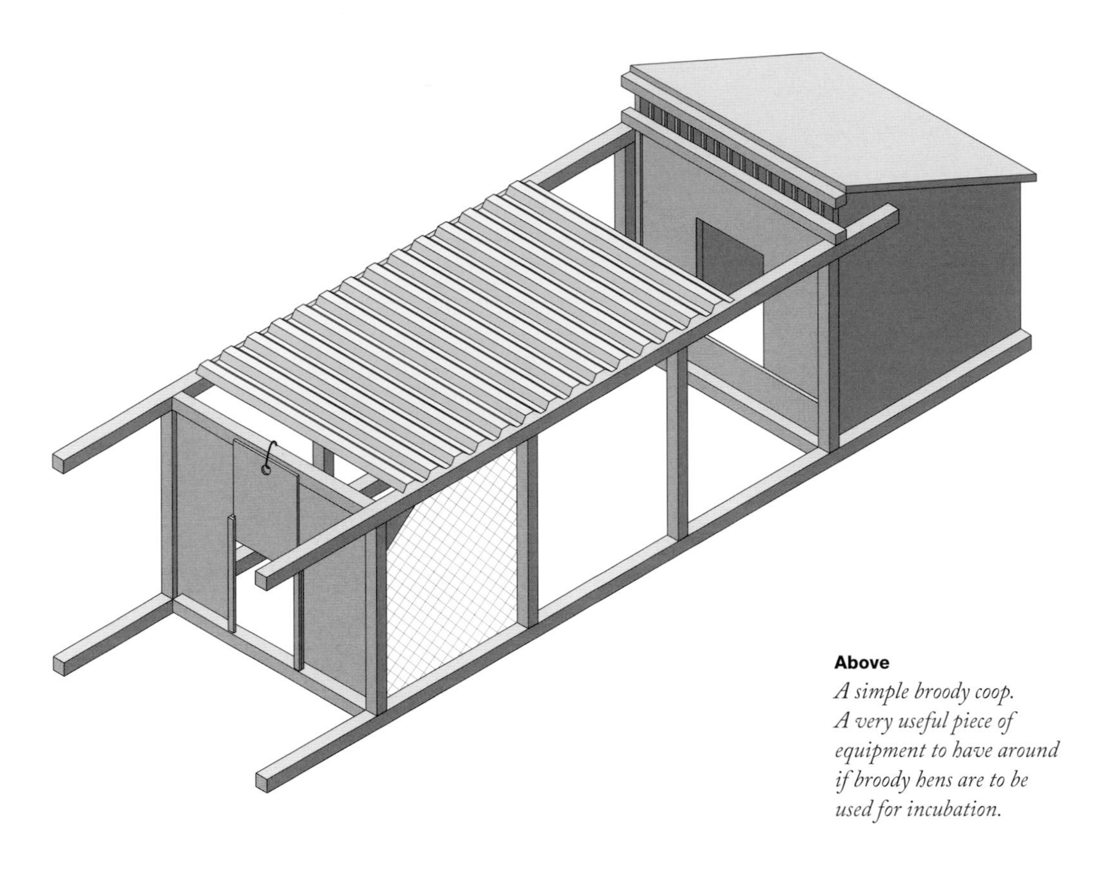

Above
A simple broody coop. A very useful piece of equipment to have around if broody hens are to be used for incubation.

broody again, repeat the process but extend it for a few days. Eventually, she will cool off and lose the desire to nest.

USING A BROODY HEN

Possibly the simplest way to incubate and hatch eggs is by using a broody hen. If she is a reliable sitter, then a good broody hen will remove a lot of the issues that can be encountered with artificial incubators (see page 32). She will provide the optimum conditions for embryonic development and will even adjust the conditions when the eggs are due to hatch. Before using a broody hen, check that she is in good condition. She should be in good health, of a suitable weight for the breed, and thoroughly checked (and treated) for external and internal parasites. She will be required to sit in the same location for 21 days, during which time she will leave the nest only infrequently, so it is important that she is fit and clean at the outset.

Broody Hen Accommodation

Some hens will go broody in the main coop surrounded by other birds. This is not ideal, and the hen should be moved to a broody coop or small animal hutch if at all possible, located

The broody poop

This requires a mention, as it can come as quite a surprise when first encountered. In essence, the broody poop is the waste passed by a hen that is currently in a state of broodiness. It is possible that she has not left the nest for a day or two, so she will empty her bowels all at once rather than frequently throughout the day.

somewhere relatively dark and quiet, and given her own supply of food and water. Leaving a broody hen with the rest of the flock runs the risk that others will go broody and join her in the nest, or that they will lay in the same nestbox, resulting in the eggs set for incubation becoming mixed up with those that were not.

If you can't move the broody hen to separate accommodations to incubate her eggs, ensure that she is settled in a dog crate or similar enclosure. This should exclude other hens but be set at a level off the ground such that any chicks that do hatch can easily return to it should they stray from the nest.

Setting a Broody Hen

Once a suitable broody hen has been identified and checked, and her accommodations have been prepared, she will need to be set on viable eggs. These eggs should be marked with an indelible ink pen to ensure that there is no confusion with any other eggs that appear in the nest. Do not use a pencil, as these markings will rub off after a few days of the hen sitting on the eggs.

In order to minimize the chances of the hen rejecting the eggs, they should be placed in the nest in the evening or, ideally, after dark. The number of eggs you include depends on their size and the size of the bird, but as a guide, an average-sized domestic hen can incubate a clutch of 10–12 average-sized eggs. Folklore suggests that an odd number of eggs should be set in to ensure good luck with the hatch.

Managing a Broody Hen

The broody hen requires little management. She will leave the nest when required to pass waste, and briefly to feed and drink. She may lose some feathers, and it is not unusual for her to pluck feathers from her belly and breast to create a brood patch, enabling a more efficient transfer of heat to the eggs. Ensure that she has constant access to food and water, and, if she is being kept in a more confined broody coop, be sure to clean out the broody poop (see box) when it appears.

There are occasions when a broody hen may abandon her nest and not return. This could be due to a number of reasons, such as disturbance by predators, excessive external parasite infestation, constant disturbance by the keeper or by other chickens in the flock, or simply because the hen is either inexperienced or of a breed that is not known for sitting.

The viability of eggs in an abandoned nest depends upon the external temperature, how quickly the eggs have cooled, and their stage of development. In general, the closer the eggs are to their hatch date, the more likely they are to survive a period of non-incubation. Should a hen abandon a nest, the keeper will need either to move the eggs to an incubator (see page 32) or place them under another broody hen that is not currently incubating a clutch or has only recently been set on eggs. If the latter, the newly set eggs should be removed and replaced with the rescued clutch.

Egg Myths

Over the years, a number of common myths have arisen around eggs.

Myth: You need a cockerel in order for the hens to lay eggs, but fertilized eggs shouldn't be eaten.

False: A hen or flock of hens will quite happily lay eggs throughout the laying season whether a cockerel is present or not. Introducing a cockerel will potentially result in him treading the hens and consequently fertilizing the eggs. However, this does not mean the eggs can't be eaten, as the actual development of the eggs would occur only if they were subject to the correct incubation conditions.

Myth: It's essential to turn eggs during incubation for successful hatching.

False: If you don't turn eggs during incubation, some will hatch, but if you do turn them, the hatch rate will be significantly higher. Turning the egg prevents the embryo from sticking to the side of the shell, allowing the chick to move into the hatching position when the egg reaches term. It also ensures that an even temperature is achieved within the egg, and it improves and refreshes the contact between the embryonic membrane and the nutrient-rich albumen within the egg.

Myth: The sex of a hatchling can be determined from the outward appearance, size, and weight of the egg containing it.

False: Despite various claims, inventions, hearsay, and handed-down tales, this is not as yet possible. Since the early 1900s, researchers around the world have studied and tested various hypotheses and have found no statistically significant way of being able to define the sex of the hatchling by observing the characteristics of the egg itself.

Myth: If a broody hen is away from the nest for more than a few minutes or if there is a power outage and the incubated eggs begin to chill, then the embryos will die.

False: In the event of a power outage, switching the incubator off, or a broody hen accidentally being locked out of her nestbox for a long period of time, embryonic development does slow down, but the embryo generally only dies if the time period exceeds 15 hours or the weather conditions are extremely hot or cold. Such breaks in incubation will delay the hatch and may result in slightly reduced hatch rates but will not necessarily result in complete failure.

Myth: Brown eggs are better for you than white eggs

False: The color of the eggshell is simply linked to the breed of chicken that produced it, while the quality and flavor of the egg are dictated by the diet of the hen that laid it. The myth that brown eggs are healthier came into being because the chicken breeds favored by commercial producers (particularly in the US) invariably laid white eggs. In most cases, these commercially produced eggs came from cage or barn systems where the diet of the birds was strictly controlled. As such, any tinted or brown egg would more than likely have been laid by a small, probably free-ranging, flock on a farm. The diet of these hens consequently would have contained more varied and flavor-enhancing foodstuffs foraged from around the farm, the result being an egg that probably tasted better but just happened to be in a brown shell.

Myth: The stronger the color of the eggshell, the better the hen that that laid it.

False: Good coloring of eggs in, for example, Welsummers or Marans is most visible at the beginning of the laying season and is valuable when it comes to exhibiting the eggs. However, as the hen lays more and more eggs during the season, her ability to color them so intensely diminishes. The process is similar to an inkjet printer, with the start of the season being equivalent to a new cartridge that slowly runs down over the year. A hen laying solid brown eggs toward the end of the laying season therefore hasn't laid many eggs through the season.

Myth: It's possible for a hen to change sex and crow like a cockerel.

False: Occasionally, a hen that has been laying eggs will suddenly appear to become a cock bird. She will no longer lay eggs, her comb and wattles will develop, her feathering will become more male in appearance and structure, and she will even begin to crow. She is, however, still a she. She has only phenotypically transitioned into a male, while genetically she remains female.

The reason for this is usually an environmental stress or illness such as a tumor, problems with the adrenal gland, or an ovarian cyst. Not all hens develop both ovaries during their embryonic stages, and instead have one developed ovary while the other remains as a regressed male gonad. In the event that the developed ovary becomes damaged and ceases to function, the gonad can take over, and the subsequent increase in male hormones causes the hen to develop male characteristics. She will, however, remain female but will not be fertile.

CHAPTER TEN

Diet & Health

For non-commercial setups, there are no hard-and-fast rules stating precisely what each chicken's daily ration of feed should be. It will vary depending on the time of year, the weather, the temperature, the rate of lay, the bird's age and weight, the amount of effort it expends during its day, and the nutritional value of the feed it is receiving. This doesn't mean that you shouldn't try to establish an average range of daily feed consumption for your stock. This will enable you to budget correctly and also spot any issues with feed intake that could signal that a flock member may be ill, broody, or otherwise not its usual self.

Commercial Feeds

Most keepers feed their flock with a commercially manufactured feed, as this ensures that they receive the correct balance of the nutrients they require (see pages 49–53). This feed usually comes in the form of mash or compound pellets. Nowadays, "mash" is the term used to describe a commercially manufactured powdery feed that needs mixing with water before giving to the chickens. The keeper needs to mix up the mash each day; once mixed with water, it will go stale quickly. Pellets are simply the same feed but pressed into a pellet shape, making them easier to manage and removing the need to mix up feed on a daily basis. They also last longer than mash mix once out of the sack and hence are more economically viable for the small-flock owner.

Below
Getting the balance right in a hen feed is essential if the hens are to perform well.

Supplements

While formulated commercial diets provide all the nutrients your hens need, a few supplements will further improve their health and well-being.

OYSTER SHELL

Oyster shell is often given to laying hens as a free-choice supplement to enhance eggshell quality, especially as they get older. An additional benefit of providing oyster shell over simply increasing the limestone content of the feed is that the large size of the particles means they dissolve slowly in the gut, providing a slow release of the minerals calcium and phosphorus. Interestingly, hens that require more calcium in their diet will sense this need and seek out and consume oyster shell. It should be noted that the slow release of these minerals makes oyster shell an unsuitable supplement for birds exhibiting a lack of calcium; instead, these birds should be given a fast-acting liquid calcium supplement.

SCRATCH FEEDS

Scratch feed is generally a mixture of any number of cereal grains—often cracked corn and wheat—and is fed by scattering it on the ground so that the hens can scratch around for it as a treat and as exercise. Scratching for food is a leftover from the natural behavior of the chicken's jungle-fowl ancestors, which searched through the debris on the jungle floor for insects and seeds. Scratch should not be considered a feed but should be used only as a treat. Consumption of large amounts of scratch can put a chicken's daily intake severely out of balance and over the long run reduce its productivity.

It is always useful to provide the flock with a scratch feed an hour or two before the birds go in to roost. This not only provides them with a full crop before roosting (particularly useful during the colder winter nights), but provides activity and stimulus, encourages flock cohesion, and keeps their toenails down.

The other advantage of getting your flock used to a daily scratch feed is that the birds will respond to you bringing it to them no matter what the time of day. If the scratch feed is provided within the run, you can attract free-ranging birds back to a fixed area at any point of the day. Alternatively, providing a small quantity in the coop will bring them indoors. Both scenarios are particularly helpful if you need to catch and handle the birds during the day or if you need to confine them earlier than their normal roost time.

GRIT

Grit is required by chickens to aid in digestion. When chickens eat, they swallow everything whole, since they don't have teeth for grinding. Foraging chicken often consume whole grains or insects with hard exoskeletons that require

grinding to be digested. When the grit is swallowed, it is lodged in the gizzard, the muscular stomach of birds, where muscle contractions grind the feed against the hard particles.

Grit comes in a variety of sizes—it is important that the size of the grit provided is appropriate for the age and breed of the chicken. Some grits contain additional trace minerals to supplement the diet. Grits should be freely available to the birds and not mixed with feed, as chickens will seek it out as and when they need it.

VITAMIN & MINERAL SUPPLEMENTS

Essential vitamins are required for the health and growth of chickens but they cannot be synthesized by the birds themselves so need to be included within their food (studies have shown that chickens require all vitamins apart from vitamin C). Free-ranging birds will forage for much of their feed and, as such, will pick up a lot of the essential vitamins they require from their varied diet. Confined birds, however, will be reliant upon the keeper and the quality of the feed provided. Most commercial feeds contain sufficient daily allowances of the vitamins chickens need, but the vitamin content will begin to degrade after the "best before" date. If your chickens aren't free-range, then it is important to ensure they have sufficient vitamins in their diet. One way of doing this is to give them multivitamins.

There are many varieties of multivitamins for chickens on the market, from pleasant-smelling powders that can be mixed in with the feed to liquid supplements that can be added to the drinking water. Different keepers have different preferences regarding which to use, and these are as much based on the method of application as on the actual ingredients. However, most keepers agree that these natural vitamins and minerals give the birds a boost during their molt, in the spring and winter, at times of stress, and in particular before exhibiting.

Right
Grit is required by chickens to aid the digestion of food.

PASTURE

There has been a resurgence in allowing chickens to forage on pasture. As a feedstuff for chickens, pasture itself is poor, as the birds can't digest cellulose, the primary component of grasses. However, they do enjoy picking out the insects that pasture attracts and the seeds the grasses produce at certain times of the year. As the grass begins to grow in the early spring, the insect population provides an excellent protein-rich food source for the foraging birds. However, this resource is limited over time as the insect population declines. In the summer, grasses that are allowed to mature produce seeds that can provide some nutrition for chickens. As late summer turns to fall, the seeds are gone and the insect population declines to the point where it is virtually nonexistent. Cold, wet, and snowy winter conditions eliminate pasture grasses as a feed for chickens. Then it's back to spring for another cycle.

Therefore, on an annual basis, little feed value is obtained by allowing chickens to forage on pasture. Rotating the birds on paddocks, keeping the numbers low enough so that the insect population is not reduced too quickly, and planting pasture grasses that attract insects and increase seed production will enhance their value as a feed supplement.

Below
Allowing access to fresh pasture can mean access to various seeds and insects for the hens.

Vitamin & Mineral Deficiencies

Fat-soluble vitamins

These are stored within the body tissue and include vitamins A, D, E, and K. If chickens or chicks are deficient in fat-soluble vitamins, the following symptoms can be evident:

Vitamin A	decreased egg production, weakness, lack of growth in chicks
Vitamin D	thin-shelled eggs, reduced egg production, slowed growth, chick rickets
Vitamin E	decreased fertility and hatchability
Vitamin K	slow blood clotting

Water-soluble vitamins

These highly soluble vitamins require daily replenishment; they include vitamins B1, B2, B3, B5, B7, B9, B12, and choline. If chickens or chicks are deficient in water-soluble vitamins, the following symptoms can be evident:

Vitamin B1 (thiamine)	loss of appetite and death
Vitamin B2 (riboflavin)	poor egg production, curled toes, poor growth rates in chicks
Vitamin B3 (niacin)	bowed legs, inflammation of the tongue and mouth cavity
Vitamin B5 (pantothenic acid)	dermatitis and lesions on the mouth and feet
Choline	decreased egg production, poor chick growth
Vitamin B7 (biotin)	dermatitis around the eyes and beak
Vitamin B9 (folic acid)	poor egg production, poor growth and feathering in chicks, anemia
Vitamin B12	anemia, poor growth, embryonic mortality

Essential minerals

Various essential minerals are also required for the health and well-being of your chickens. If there are deficiencies, the following symptoms can be evident:

Calcium	poor eggshell quality and hatchability, rickets
Phosphorus	poor eggshell quality and hatchability, rickets
Magnesium	sudden death within a flock
Iron	anemia
Copper	anemia
Iodine	enlarged thyroid (goiter)
Zinc	poor feathering
Cobalt	slow growth and increased mortality in chicks, reduced hatchability

Homemade electrolyte solution

Take 9.5 pints (4.55 liters) of water and add one tablespoon of sugar (sucrose), one teaspoon of baking soda (sodium bicarbonate), one teaspoon of salt (sodium chloride), and half a teaspoon of salt substitute (potassium chloride). Mix until dissolved and then offer to the chicken in place of normal drinking water for five hours before replacing with normal drinking water. Repeat this process for five to seven days, and then reinstate normal drinking water altogether. By this point, the chicken should be showing a marked recovery.

If a chicken is suffering dehydration, for example due to an extended bout of diarrhea, or is recovering from a heavy worm infestation, then essential minerals can be leached from the body. The health and well-being of a chicken can be dependent upon these minerals, so it is important that they are replaced in order to enable a rapid recovery. This can be achieved by providing the chicken with an electrolyte mix.

Pre-mixed electrolyte powders and solutions are available off the shelf, but it is possible to mix your own from ingredients found in the home (see box). This is especially useful if the feed store is some distance away, or the need arises outside of normal business hours.

APPLE CIDER VINEGAR (ACV)

This supplement can be added to the chickens' drinking water, but because it is acidic, note that it should not be used in metal waterers, or they will corrode. ACV is used as a general health supplement for other animals and for humans, but be sure to use the unrefined product for livestock, which can be purchased from agricultural merchants, as opposed to the ACV available for people.

As a dietary supplement, ACV can help control intestinal pests, such as worms, by marginally changing the pH of the digestive system. The dosage rate is 5 teaspoons of ACV to 2 pints (1 liter) of drinking water, which should be given to the chickens for a period of 10–14 days. This can be followed by maintenance periods at the same concentration of either two days a week (the weekend, for example), or one week per month.

Right
Oregano is also known as wild marjoram. It has purple flowers and olive-green leaves. Its dietary benefits make it a useful addition to your run.

OREGANO

Oregano is another useful dietary supplement, with antibacterial, antifungal, antiparasitic, and antioxidative properties; it also helps boost the immune system of chickens. Oregano oil extracts can be purchased, but hanging a bunch of fresh oregano in the run not only provides the herbal health benefits but also gives your chickens some greens and entertainment.

GARLIC

The health benefits of garlic are widely known and written about, again not only in humans, but in animals such as horses. Garlic is also very beneficial for chickens and, if not used to excess, doesn't taint the flavor of eggs.

Garlic can give the chicken's immune system a boost in much the same way as it does for humans, and there is evidence to suggest it reduces infestations of internal and external parasites, particularly of the blood-sucking

variety. It has also been proven to reduce the odor of chicken manure, so, all in all, it is a handy supplement to the diet.

Garlic granules can be purchased at most feed stores, although they are usually in the equine rather than the poultry section. These can be scattered into the chicken feed, and most chickens seem to like the taste, picking it out of the feed when they sense it's there. There are also products on the market that contain concentrated garlic extract (allicin), which can be used both as a preventative and a treatment for worms and other parasites that affect chickens. These products are usually added to the chickens' drinking water.

A homemade alternative is to add a clove or two of garlic to your chickens' waterer. It will gradually become mushy and release its oils into the water for the chickens to ingest. Once the waterer is empty, don't throw the clove away; hand it to the nearest chicken, which will enthusiastically consume it. Again, the strength of flavor in the clove will have dissipated so as not to taint the eggs of that bird.

COD LIVER OIL

In much the same way that garlic has been known to provide health benefits in animals, so has cod-liver oil. It is rich in vitamins A and D, and is a valuable general aid to feather conditioning, especially during periods of molt. Like garlic, if used in moderation, it will not taint the taste of eggs.

Cod-liver oil is also particularly useful in helping powder-based medications, such as worming powder, stick to the chickens' feed, ensuring that the medication is consumed and doesn't simply fall to the bottom of the feeder. Simply measure out the required powder and sufficient feed to ensure that all birds being treated will receive their daily ration, and then mix the two together with two teaspoons of cod-liver oil per 1 lb. (450 g) of feed.

Catching, Carrying, & Crop Reading

Spend enough time around a flock of chickens and it will soon become apparent that there is a significant amount of husbandry by eye that is required. A good stockperson will be able to cast a glance across the animals and spot anything that seems out of line. This is a skill that is particularly worth acquiring when it comes to a flock of chickens, as the birds can disguise ailments that might identify them as being weak or vulnerable to predation.

A hunched or drooped stance, a discolored comb or droppings, poor feather condition, or lack of alertness can all be indicators of illness and should be investigated by catching the hen. The simplest time to do this is when a bird is roosting and is perched. It can easily be lifted off and handled. If you need to catch the bird during the day, try to corral it into a corner.

Once the chicken is in hand, trap its wings against the sides of its body. With a tame bird, this can be done simply with your thumb and little finger as the bird's undercarriage sits in the palm of your hand. If the bird is not tame and is struggling to escape, hold it with one of the wings to your chest and one hand over the other wing. Your second hand should then be used to hold the legs firmly together (but not too tightly). If the bird continues to struggle, place a cloth over its eyes to help it settle.

CROP READING

Whenever you are holding a chicken, it is always worth checking the bird over for parasites and assessing its overall condition as you did when you purchased your birds (see page 133). Crop reading is also worthwhile. The crop is the first section of a chicken's digestive system reached after food has entered the mouth and traveled down the esophagus. It sits beneath the neck of the chicken and toward the front of the breast, and can easily be felt by the keeper when the bird is in the hand.

"Reading" the crop by gently feeling its condition means you can perform a basic health check on the bird. Normally a crop will contain food (unless the bird has yet to eat that day) and will feel like a slightly soft ball when squeezed gently. If the bird has not eaten or is off its food, the crop should feel empty and almost absent. If the crop feels solid and hard, the bird could have an impacted crop, whereas if it feels watery and squishy (and the bird's breath smells strong), the bird could have sour crop.

Being familiar with these basic crop conditions will help significantly in managing the welfare of individual birds. If you think a member of your flock has an impacted or sour crop, or you are concerned about its condition, refer to the table opposite and overleaf, and contact your vet as soon as possible.

Common Diseases & Health Problems

The table below gives a brief summary of the most common diseases and health problems affecting chickens kept on a small scale. If you have any concerns over the health of a bird or birds in your flock, get in touch with your local vet for a professional diagnosis.

Common Diseases

Illness/disease	Symptoms & signs	Causes	Management
Coccidiosis (see page 169)	Young stock listless, head sunk in neck, possibly blood in feces	*Coccidia* parasite	Keep litter dry and clean. Use a water sanitizer such as apple cider vinegar in a plastic waterer (see page 164), clean run and coop with parasitic cleanser, seek vet attention for medication.
Parasitic worms (see page 169)	Listless, greenish diarrhea, gaping	Parasitic worms, many species	Regular worming with an appropriate wormer (frequency dependent upon stock density and land conditions). Eliminate standing water.
Enteritis	Brown diarrhea, slow growth, slow feathering	*Escherichia coli*, chilling, heat stress, poor-quality feed, unhygienic litter conditions	Improve feed quality, remove stress factors, keep litter clean and dry.
Bacilliary white diarrhea (BWD)	White diarrhea, thirst, sudden death	*Salmonella pullorum*	Have your vet carry out a blood test to establish which birds are carriers; cull carriers.
Vent gleet	Foul-smelling, scabby vent, fecal encrusting	*Herpes* virus	Cull affected birds.
Avian tuberculosis	Weight loss despite being alert and feeding	*Mycobacterium avium*	Cull affected birds. Wild birds tend to be carriers so avoid contact between domestic and wild flocks.
Mycoplasmosis	Sneezing, nasal discharge, foam in eye, rattly breathing, swollen sinus	Mycoplasma	Reduce stress, clean coop, ensure nutrition is good quality. Consult your local vet.

Illness/disease	Symptoms & signs	Causes	Management
Bumblefoot	Limping, swelling on foot pad	*Staphylococcus* bacteria	Invasive surgery is required but has only limited success. Ensure proper perches that are not too high. Remove sharp objects and eliminate wet litter.
Marek's disease	Paralysis down one side, wing, and leg.	*Herpes* virus	Cull affected birds. Consider purchase of vaccinated chicks.
Heart and/or circulation problem	Red comb and wattles darkened, bird listless	Many, including old age and disease	Provide fresh feed and fresh water daily.
Sour crop	Pendulous, soft crop	Yeast overgrowth	Induce vomiting to empty crop. Treatment with anti-fungal agent is sometimes recommended.
Impacted crop	Pendulous, hard crop	Fibrous grass	Seek vet attention, as surgery may be required. Keep grass short in run area.
Aspergillosis	Respiratory problems	Fungal infection	Cull affected birds. Avoid use of bark litter or hay. Keep ground free from damp litter.
Infectious bronchitis	Respiratory problems	Coronavirus	Consult your local vet.
Egg bound	Agitated behavior, bearing down, listless	Egg stuck in oviduct	Leave hen to lay in a warm, dark place.
Internal layer, egg peritonitis	Agitated behavior, bearing down, listless	Yolk in the abdomen	Usually results in death. Culling advised.
Red mite (see page 170)	Anemia, occasionally death in severe infestations	Red mite (*Dermanyssus gallinae*)	Fill cracks and gaps in henhouse and clean it thoroughly and regularly to break parasite's life cycle.
Northern fowl mite (see page 171)	Sunken stance, greasy-appearing feathers, mites visible at base of tail, death may result if not treated	Northern fowl mite (*Ornithonyssus sylviarum*)	Wash birds with medicated shampoo as soon as infestation is spotted.
Scaly leg mite (see page 171)	Scabby legs	Scaly leg mite (*Knemidocoptes mutans*)	Wash the affected areas and rub on petroleum jelly to act as a barrier.
Lice (see page 170)	Poor feather condition	Approximately 50 species of lice	Application of insecticidal louse powder.

INTERNAL PARASITES

Coccidiosis

Coccidiosis is a parasite-induced illness that can be debilitating and fatal to chickens. Young growing stock will be exposed to the parasite (*Coccidia protozoan*) at some point their lives, and they will usually develop an immunity to it, either because the exposure is at a low level or because they have survived the resulting illness.

Artificially reared birds can be particularly at risk of contracting the condition when they are first put outdoors. The outward signs of the disease are a hunched, fluffed-up bird with drooped wings; it can also present itself as blood in the droppings of the bird, although this doesn't always occur. The condition can be confirmed by having fecal samples tested by a vet. A rapid response to tackling the problem is required, as it can very quickly spread throughout the whole flock when the infected droppings come into contact with uninfected birds, potentially wiping out a batch of young birds within days. Keeping ground litter dry and clean can help prevent its occurrence, but cleaning the coop and run with a suitable parasitic cleanser and using an oral medication supplied by a vet will be required in the event of an outbreak.

The life cycle of coccidiosis

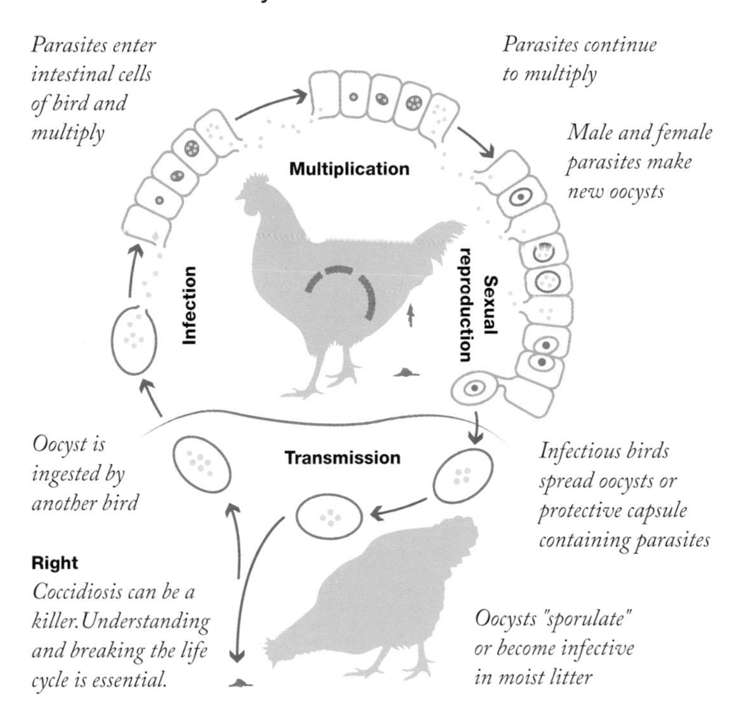

Right
Coccidiosis can be a killer. Understanding and breaking the life cycle is essential.

Parasitic Worms

Chickens are no different from most other animals in that they are susceptible to infestation by parasitic worms that have evolved to exploit them. The two groups that are significant to chickens are flatworms (cestodes and trematodes) and roundworms (nematodes). Both of these types of worms live within chickens, feeding off their host and reproducing while protected from the outside world.

An infestation of worms can reduce the nutrient uptake of the chicken and disrupt its immune system by creating an additional burden. In most cases, chickens will develop a level of resistance to worm infection, though if the flock is kept on the same area of land, the worm count can increase significantly and the flock may become heavily infested.

Regular worming can be performed using over-the-counter poultry wormers if movement to fresh ground is limited. Additionally, worming in the spring and again in the fall is advisable in order to reduce the risk of other diseases compromising the flock.

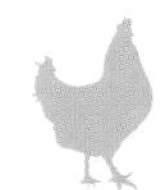

EXTERNAL PARASITES

Lice

As with most creatures, chickens can be hosts to a wide variety of lice. There are around 50 different types of lice that can affect chickens, with different species exploiting different parts of the body. In general, they tend to be small parasites that scuttle around on the skin of the chickens and hence can be quite difficult to spot. It is therefore important that the keeper performs a thorough inspection of the bird when it is in the hand, paying particular attention to the base of the feather shafts, as this is frequently where the lice eggs (or nits) can be found.

In most cases, lice tend to be more of an irritation than a direct life-threatening risk, although, if not dealt with quickly and effectively, heavy infestations can cause extreme stress and possible fatality in a flock.

As the lice life cycle is invariably spent entirely on the host, treatment is usually applied directly to the chicken with the application of a suitable insecticidal louse powder. Ensuring that the birds also have a dust-baths will also help them tackle the problem themselves, especially if the louse powder is added to the bath.

> **Tip**
>
> Keeping grass short leaves few hiding places for smaller microscopic pests such as parasitic worm eggs, which don't survive well when exposed to UV light rays. It also means the area is a little more exposed and less welcoming to larger predators. Don't denude the run, but keeping some of the grass and scrub in check will help control pests and predators.

Red Mite

The red mite (*Dermanyssus gallinae*) is a pest of the summer months, and, while it may not be active during winter weather, it has an incredible ability to survive long periods without feeding.

The parasite takes up residence within the chicken coop and usually completes its life cycle off the bird, living in crevices or cracks in the structure of the henhouse. It gets its sustenance by sucking the blood from the chickens as they roost at night, before returning to the crack from which it came. Red mites can breed and multiply at an incredibly rapid rate, completing their maturation from egg to egg-laying adult in a matter of days. If left unchecked, their resulting population can be so large that the quantity of blood they take from a bird overnight can result in anemia and even death. A red mite can survive more than six months without feeding, and, while a really harsh winter may kill off adult mites, the eggs they laid late in the summer will live on, ready to hatch at the first signs of warm weather.

A red mite infestation is notoriously difficult to get rid of once it has taken hold of a coop. A regimented cleaning and treatment program is required over a number of weeks in order to try to break the mite's life cycle. It is far better to act quickly as soon as the parasite's presence has been identified.

A simple method of checking for red mites within a coop is to place a bunch of drinking straws at floor level in the corner. Each week, remove the straws and blow through them into an empty jar. If mites are present in the henhouse, they will probably take shelter in the drinking straws and will be seen crawling around in the bottom of the jar.

Alternatively, place a piece of corrugated cardboard at floor level, with the corrugated side face down. If red mites are present, they will take up residence in the cavities of the cardboard.

Northern Fowl Mite

Another mite to be aware of is the northern fowl mite (*Ornithonyssus sylviarum*). Chickens infested with the parasite tend to have a sunken stance and a greasy appearance to their feathers. Picking up the chicken and parting the feathers at the base of the tail will usually reveal the mites crawling over the skin. Beware—they will also crawl onto your hand, seeking you out as a potential host, and while this will cause only a minor irritation, it is unpleasant and runs the risk that the mite is transferred to another bird or flock. Northern fowl mites become far more active during winter because they prefer cooler climates, and they usually infest male birds in preference to females, although both sexes are susceptible to attack.

Once northern fowl mites find a suitable host bird, they will multiply at an incredible rate and, like the red mite, they are blood suckers. Unlike the red mite, however, these parasites complete their entire life cycle on the birds, are far more aggressive, and feed around the clock. The greasy look of the feathers is caused by their fecal deposits and is usually a strong visual indicator of an infestation. Treatment needs to be rapid and is usually achieved by washing the bird with a medicated shampoo. The rapid reproduction and voracious feeding habits of the mites mean they are capable of killing a bird within a matter of days if the infestation isn't dealt with immediately.

Scaly Leg Mite

A less lethal but equally irritating poultry pest is the scaly leg mite (*Knemidocoptes mutans*). An infestation of these mites is difficult to identify until it is well underway. The tiny flat-bodied parasite is not usually seen, but the consequences of its presence are easily visible. The mite lives under the scales on the legs and feet of chickens, and over a period of time the scales begin to lift and crust over. This can cause the legs to thicken and become scabby in appearance, which is usually the first indicator of a scaly-mite problem. As with other mites, quick action is required. Both the infested bird and any others it has come into contact with should be treated because the mite is primarily transferred through direct contact between chickens.

Left
Red mite, the poultry-keeper's nemesis, can kill if left unchecked.

The Poultry First-Aid Kit

Every poultry-keeper should have a first-aid kit containing a selection of items, some of which will probably be needed at least once a week. It is useful to keep them together in a bag or box so they are handy, which also makes it easier for anyone who cares for your birds while you are away.

FLASHLIGHT *(1)*
It's often easier to handle and treat chickens after they have gone to roost, as they then tend to be much calmer and easier to handle. This does, however, mean you will be working in the dark, so you will need a flashlight. Investing in a good-quality flashlight, particularly one with adjustable beam strength, will mean you have both hands free to deal with the chickens.

SCISSORS *(2)*
A strong, sharp pair of scissors is needed for cutting string and bandages. Most of all, scissors are needed to clip flight feathers if you have a flighty chicken in your flock that keeps jumping out of the pen (see page 137).

TOENAIL CLIPPERS AND NAIL FILE *(3)*
Most chickens will keep their toenails worn down by scratching around, but birds kept indoors or on soft ground may require a bit of a pedicure. In addition, some cockerels do need to have their spurs attended to (see box on page 138).

PLIERS OR WIRE-CUTTERS *(4)*
These are not only useful for emergency fence repairs, but are also ideal for quickly removing plastic leg rings (see below).

LEG RINGS *(5)*
Having a range of different-colored and different-sized leg rings on hand means you can quickly and easily "mark" an individual chicken. This is particularly useful if you are administering treatments to the flock and need to separate those that have been treated from those yet to be dealt with.

FEEDING SYRINGES *(6)*
A collection of different-sized feeding syringes is essential when you need to administer fluids such as medicines down the throat of a chicken.

DISPOSABLE LATEX GLOVES *(7)*
Gloves are not needed often, but when it comes to vent-related problems such as a prolapse or vent gleet (see page 167), they can make the task much easier for the keeper (and probably more comfortable for the chicken, too).

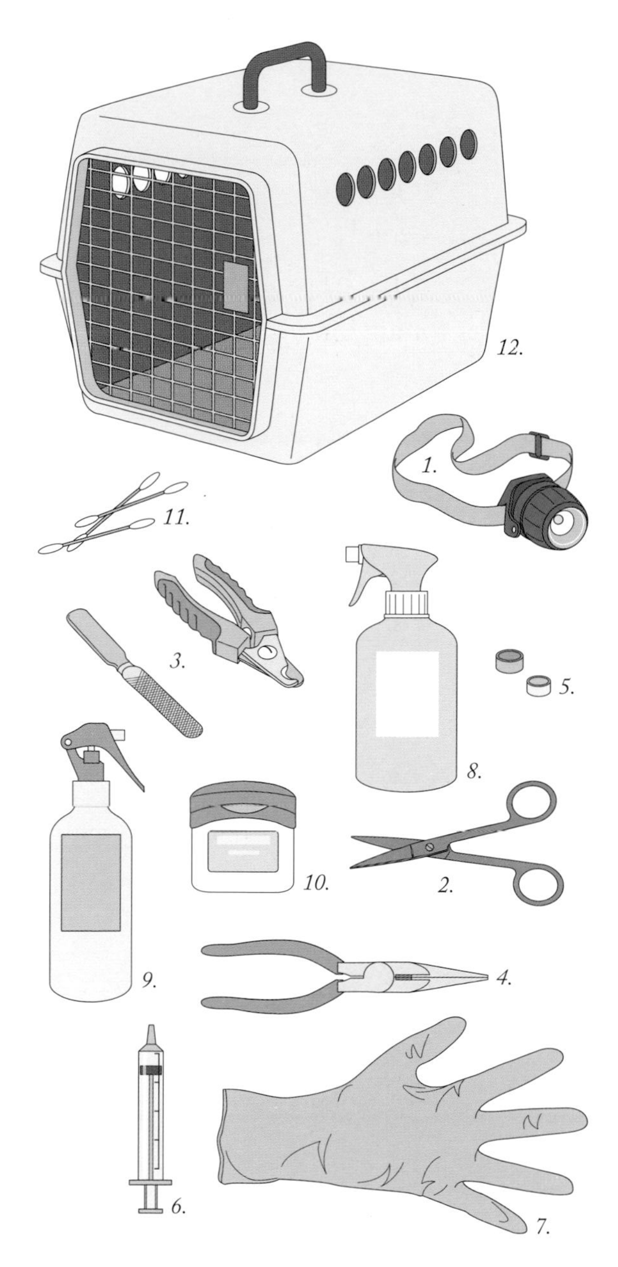

ANTISEPTIC SPRAY *(8)*
A veterinary antiseptic spray is ideal for treating minor wounds and can also double up as a deterrent spray on feathers in minor cases of feather pecking.

GENTIAN VIOLET SPRAY *(9)*
This spray, popular in the UK, works in much the same way as the veterinary antiseptic spray, but it has the advantage of being visible. This means that it can also be used to mark birds quickly and temporarily either after treatment or for further selection. Don't use it on chickens you intend to exhibit as it can be difficult to remove fully.

PETROLEUM JELLY *(10)*
Not only does this serve well as a lubricant for sticky hinges and locks on the henhouse, but it can also be applied carefully to the combs of birds during extremely cold weather to reduce the risk of frostbite, and to dry patches of skin on their faces or legs. It is also handy when treating scaly leg mite on chickens should they become infested with the parasite (see page 171).

COTTON SWABS *(11)*
Cotton swabs are useful for delicate tasks such as cleaning around the eyes or nasal passages of birds.

PET CARRIER OR DOG CRATE *(12)*
You can never have too many pet carriers for transporting or quarantining your chickens. Plastic dog or cat carriers are ideal for single or even small numbers of chickens, but be sure to disinfect them thoroughly after each use to avoid any possible transferral of pests or diseases to other chickens.

Finally, make sure your vet's telephone number is readily available at all times. While you may not necessarily need it, it could prove invaluable if friends or neighbors look after your stock while you are away.

Glossary

AIR CELL The pocket of air within an egg, used by the chick to take its first breath before hatching. In eggs for eating, the size of the air cell is used as a measure of quality—the smaller the air cell, the fresher the egg and the higher its quality rating.

ALBUMEN The egg white, comprising almost exclusively protein and water.

BANTAM A catch-all term for a diminutive chicken. Some bantams are simply small versions of a large-fowl breed, while others are true bantams—breeds that exist only in the miniature form and have no large-fowl counterpart.

BROODY A hen that has decided to sit on a nest. Usually the nest will contain a clutch of eggs, but it is not unusual for the hen to brood an empty nest or one that contains dummy eggs (ceramic eggs used to encourage broody behavior).

CAPON A cockerel that has been castrated in order to make it grow larger and plumper for the table. Caponized cockerels have underdeveloped combs and wattles compared with uncastrated cockerels and are said to be calmer and crow less. Caponizing can be done physically or chemically, but is illegal in some countries.

COMMERCIAL STRAIN A chicken specifically developed for egg or meat production. Such birds are invariably produced by intensive selective breeding with one objective in mind: to create the maximum amount of product at the most efficient feed conversion rate within a fixed period of time.

CROP READING Feeling the crop of a chicken to ascertain what state it is in and thus gain an indication of the health of the bird.

DUST-BATH A patch of bare earth or sand, or a container containing diatomaceous earth or wood ashes, in which chickens roll to clean their feathers and remove parasites.

HARD-FEATHERED When the feathering on a bird is tight and held close to the body and may even be absent from some areas, such as the breastbone. There is little fluff on the feathers, which are usually short, narrow, and closely webbed in appearance, with strong shafts. Such feathering predominates in the game breeds, as it harks back to the days of cock fighting, when tight feathering (and reduced comb and wattle growth) provided less opportunity for grip by an opposing bird in a fight. (See also "Soft-feathered.")

HEAVY BREED A breed in which the cock weighs more than 8 lb. (3.6 kg) and the hen more than 6 lb. (2.7 kg). (See also "Light breed.")

HERITAGE BREED A pure breed that was established prior to the mid-20th century.

INFUNDIBULUM The funnel-shaped end of the oviduct closest to the ovary.

ISTHMUS The section of the oviduct in which the inner and outer shell membranes are deposited onto the albumen.

JUNGLE FOWL Pheasants originating from Southeast Asia that are the wild ancestors of domesticated chickens.

LARGE FOWL Sometimes also referred to as "standard-sized chickens," these are the original representation of the size of the breed. While one large-fowl breed may be considerably bigger than another large-fowl breed, they are both of the original and expected size that they should be for their particular breed. Within this group, the breeds are usually split further into "light" and "heavy."

LIGHT BREED A streamlined breed in which the cock weighs up to but no more than 6 lb. (2.7 kg) and the hen up to but no more than 4.5 lb. (2 kg). Light breeds, by nature, tend to be flighty, alert, and busy birds that are not easily tamed. (See also "Heavy breed.")

MAGNUM The section of the oviduct in which the albumen is deposited around the yolk.

MEAT BREED A breed developed specifically for meat provision. The males of some breeds exceed 11 lb. (5 kg) in weight, with hens not much less, thus maximizing the potential of both sexes to be used in the kitchen.

ORNAMENTAL BREED A breed whose development was, or has become, focused purely on appearance.

OVIPARITY The characteristic of laying eggs, as opposed to giving birth to live young.

OVIPOSITION The process of laying an egg.

PHOTOPERIOD The length of time an organism is exposed to light each day. In chickens, photoperiod is a trigger for egg-laying.

POINT OF LAY (POL) The point at which a pullet first starts laying eggs, usually when it is around 18–20 weeks old.

POP-HOLE The access hole that allows chickens entry to, and exit from, a henhouse.

PULLET A female bird less than one year old.

RARE BREED Within the exhibition scene, "rare breed" refers to any breed of chicken that doesn't have its own club following. Outside of the exhibition circuit, a rare breed tends to be defined at a country, continental, or global level, and relates to the numbers of the particular breed within the specified geographical area.

SHELL GLAND The section of the oviduct in which the shell is deposited. In some sources, it is incorrectly referred to as the uterus.

SOFT-FEATHERED When the plumage is not hard-feathered; commonly refers to breeds that have loose feathering, where the fluffy appearance of the feathers and their general abundance can make the bird appear much larger than it actually is in the flesh. (See also "Hard-feathered.")

Index

Picture Credits

The publisher would like to thank the following individuals and organizations for their kind permission to reproduce the images in this book. Every effort has been made to acknowledge the images, however we apologize if there are any unintentional omissions and would be grateful if notified of any corrections that should be incorporated in future reprints or editions of this book.

All artworks by **Kate Osborne** unless stated.

Alamy/Adrian Sherratt: 40-41; Agencja Fotograficzna Caro: 20; Danita Delimont: 115; Rex May: 42; Zoo Imaging Photography: 135.

Andy Cawthray: 131

Corbis/68/Bill Boch/Ocean: 144; Christie's Images: 7.

Dover images: 14, 15.

Science Photo Library/Power and Syred: 38.

Shutterstock: 11, 17, 21, 30, 34, 39, 47, 51, 52, 54-55, 97, 99, 105, 110, 113, 117, 124, 127, 136, 146, 148, 149, 150, 157, 159, 161, 162, 165, 171.

John Woodcock: 22, 32, 43, 45, 109, 119, 137, 139, 154, 169, 173.